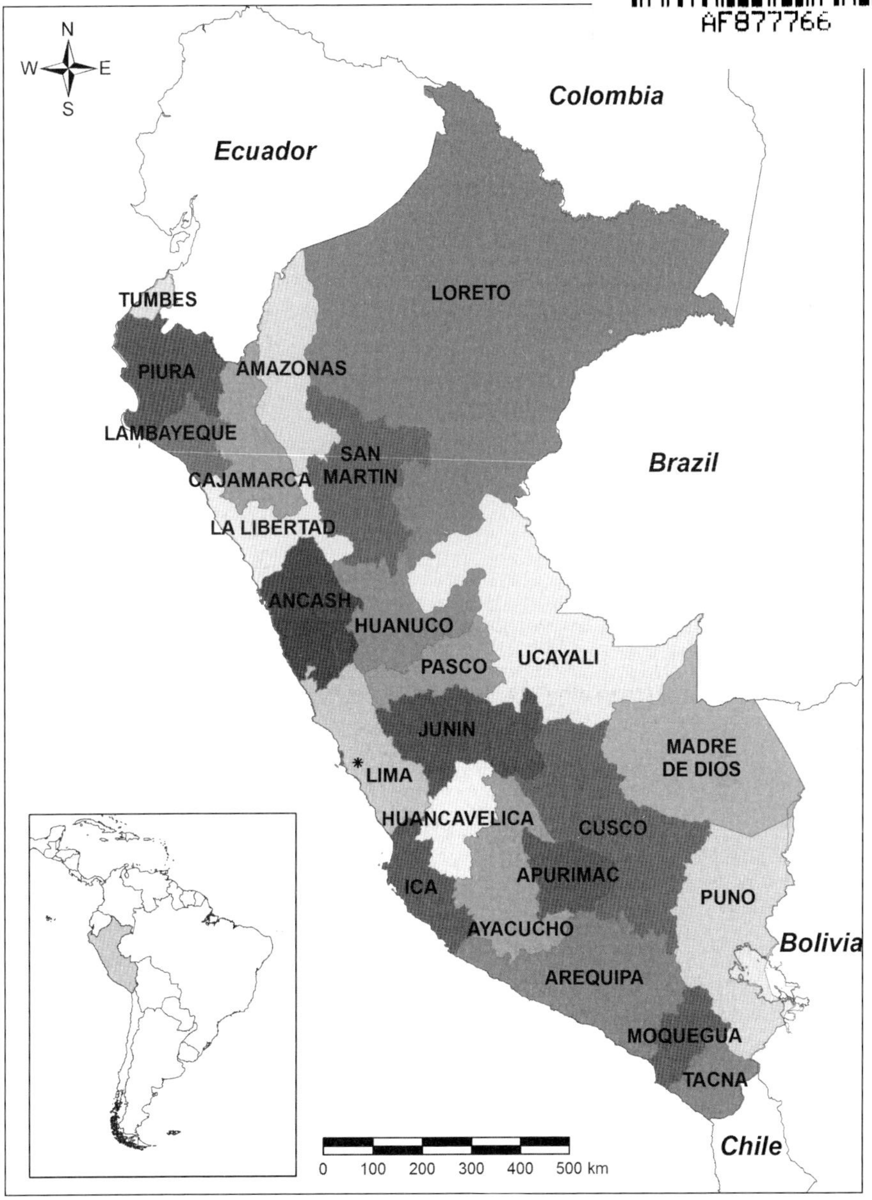

Departments

Amazonas	Huánuco	Pasco
Ancash	Ica	Piura
Apurímac	Junín	Puno
Arequipa	La Libertad	San Martín
Ayacucho	Lambayeque	Tacna
Cajamarca	Lima	Tumbes
Callao	Loreto	Ucayali
Cuzco	Madre de Dios	
Huancavelica	Moquegua	

The Mosses of Madre de Dios, Peru

The Mosses of Madre de Dios, Peru

Piers Majestyk and John P. Janovec

BRIT
PRESS
37
2011

Number 37
SBM: Sida, Botanical Miscellany
ISBN 13-978-1889878-36-2
ISSN 0883-1475

Editor: Barney Lipscomb
Botanical Research Institute of Texas
1700 University Dr.
Fort Worth, Texas 76107-3400, USA
Contributing Spanish Editor: Félix Llamas
Dpto. de Biodiversidad y Gestión Ambiental, Universidad de León
Campus de Vegazana s/n, 24071 León, Spain
Production: Rebecca Horn, rhorgraphics, rlrchorn@verizon.net
Cover Design: Christian Franklin (BRIT)

The Mosses of Madre de Dios, Peru, by Piers Majestyk and John P. Janovec

Front cover image: View from a high terrace along the Madre de Dios River overlooking a *Mauritia flexuosa* L.f. (Arecaceae) swamp (photo credit, John P. Janovec).

Back cover image: Piers Majestyk collecting mosses from the trunk of *Mauritia flexuosa* L.f. (Arecaceae) in a swamp along the Madre de Dios River (photo credit, John P. Janovec).

Map of Peru: Courtesy of Mathias Tobler.

Frontispiece image: A collage of images from Madre de Dios, Peru (photo credits: top row, left image: Antonio Vizcaino. Bottom row, middle image: Mathias Tobler. All other images in the collage: John P. Janovec).

The Mosses of Madre de Dios, Peru is published with the support from the US National Science Foundation Grant No. 0717453 to the Botanical Research Institute of Texas

Distribution of copies by:
Botanical Research Institute of Texas
1700 University Dr.
Fort Worth, Texas 76107-3400, USA
Telephone: 1 817 332 4441; Fax: 1 817 332 4112
Website: http://www.brit.org/brit-press/
Email: jbrit@brit.org

BRIT PRESS

1 September 2011

The Mosses of Madre de Dios, Peru

*A Book to Promote Botanical and Environmental Education,
Scientific Exploration, and Conservation in Madre de Dios, Peru.*

DEDICATION

We dedicate this book to Bill Buck and Scott Mori of the New York Botanical Garden, important mentors who have inspired, guided, and supported us in our studies of the neotropical flora.

APPRECIATION

The authors would like to thank the U.S. National Science Foundation (Grant No. 0717453) for funding that supported the publication of this book.

TABLE OF CONTENTS

PREFACE

The Amazonian region is the single largest phytogeographical unit recognized in the neotropics. Despite its huge size, the region has a relatively low diversity of mosses when compared to other floristic regions in the neotropics. Churchill (1998) estimates the moss diversity for the entire Amazon region at 39 families, 101 genera, and 311 species. This florula, based largely on collections from July to September 2002 in Madre de Dios, Peru, covers roughly 20% of the estimated taxa currently known in the Amazonian region. We hope that this florula will serve as a foundation for future bryological studies within and outside the area.

ACKNOWLEDGMENTS

We are grateful to the New York Botanical Garden and, in particular, Bill Buck who encouraged and supported Majestyk's field work in Peru. We thank Scott Mori of the New York Botanical Garden for his encouragement and support of Janovec's work in Peru since initial stages in 2001. We thank the Missouri Botanical Garden Press, the New York Botanical Garden Press, and the Van Eedenfonds Foundation for allowing us to use previously published illustrations in this manual. We extend our special appreciation to Barney Lipscomb and the staff of the Botanical Research Institute (BRIT) Press, as they have handled the lion's share of the final stages of bringing this book together. We thank Sy Sohmer, Director of BRIT, along with Cleve Lancaster, Pat Harrison, Will McClatchey, the Board of Directors, and staff of BRIT for institutional support and infrastructure that facilitated all phases of this project. We are especially grateful to Keri Barfield of BRIT for reviewing various versions of the manuscript, for organizing the illustrations and index, and for constant support of this and other projects. We thank Christian Franklin for designing the cover and frontispiece art, Robert George for input on requesting permission to use previously published illustrations, Justin Allison for reviewing and editing the original manuscript, Amanda Neill and Tiana Franklin Rehman for herbarium support, Jason Best for artistic input, Mathias Tobler for technical support, and Asha McElfish, Renan Valega, Jason Wells, and Pedro Centeno for logistical support. We appreciate Asuncion Cano, Betty Millan, Joaquina Alban, and Miguel Chocce of the San Marcos Herbarium of Lima, Peru, for supporting our botanical research in Peru. We are grateful to Stephen Churchill and Stephen Timme for taking the time to review the manuscript and for making many helpful suggestions for its improvement. Early stages of this work were supported by funding from the Amazon Conservation Association, the Beneficia Foundation, the Discovery Fund of Fort Worth, Texas, and the Gordon and Betty Moore Foundation. Final stages were supported by a NSF-PEET grant (#0529593) to Duke University and a NSF BIO/DEB Biotic Survey and Inventory grant (#0717453) to BRIT. Lastly, this work would not have been possible without research, collection, and export permits issued by the former Instituto de Recursos Naturales (INRENA) and the Dirección General de Flora and Fauna Silvestre (DGFFS) of Peru.

The Mosses of Madre de Dios, Peru

Bank and floodplain forest of the Madre de Dios River in southern Peru (Photo credit: John P. Janovec).

INTRODUCTION

When one thinks of a rainforest in the Amazon the first plants that come to mind are probably towering canopy trees, elegant palms, and the wandering woody vines. Bryophytes, the most ancient, living lineage of land plants and dating back more than 400 million years, are small, inconspicuous, and underappreciated. They are composed of three different lineages, including the Anthocerophyta (hornworts), Bryophyta (mosses), and Marchantiophyta (liverworts). The most diverse group of Bryophytes are the mosses, which are easily distinguished from the flat, thallus-like colonies of hornworts and liverworts by their tiny stems and leaves, and an emergent, stalked sporangium. While the diversity of mosses in an Amazonian rainforest does not compare with that of flowering plants, their abundance as clumps, cushions, or dangling colonies on soil, organic matter, rocks, stumps, trunks, and even leaf surfaces contributes significantly to the many green tones of the forest floor, understorey, and canopy. Just as mosses are overshadowed by the trees, palms, and vines in the Amazonian rainforest, so too is their importance for such things as water retention, humidity, invertebrate habitat, and germination grounds for seeds.

Here we present a guide to the mosses currently known from eastern Madre de Dios, Peru, with focus on the Los Amigos River watershed and specifically, the Los Amigos Biological Station and Conservation Concession. While our field work focused mostly on Los Amigos, we attempted to integrate specimen and species data from previous studies, such as that of Vargas (1974). Our hope is to promote awareness and interest about mosses and their diversity in the Amazon, and to provide a tool for those who desire to learn to collect and identify these fascinating members of the plant world in the Amazon, and beyond.

This work thoroughly updates and complements Majestyk and Janovec (2004), following *Seeds of the Amazon* (Cornejo & Janovec 2010) as the second book in a series of similar publications being produced based on research in Peru by the Botanical Research Institute of Texas. After summarizing some relevant background information, we present an identification key, descriptions, and illustrations for the mosses currently known from the study region.

BACKGROUND INFORMATION

The Southwestern Amazon

The Southwestern Amazon of the trinational corner of SW Brazil, NW Bolivia, and SE Peru contains what is probably the largest and least disturbed area remaining of Upper Amazonian and Lower Andean ecosystems (Foster et al. 1994). Threats to the forest occur in the form of hunting, gold mining, timber extraction, impending road construction, and slash and burn agriculture. The Department of Madre de Dios (12°–14.5°S), Peru, dominated by the Madre de Dios River basin and tributaries, is an important geopolitical region in the SW Amazon, with vegetation divided equally between lowland, premonate, and montane forest (Davis et al. 1997). Madre de Dios lies at the southwestern edge of the Amazon basin near the Andean foothills and is covered primarily by lowland tropical/subtropical moist forest (Gentry & León 1997). The region has a distinct four to five-month dry season (Jun–Sep) and receives 2000–2900 mm of rain annually (Gentry & León 1997). Madre de Dios has been recognized as a region of high diversity and holds world record numbers of birds, tabanid flies, tiger beetles, damselflies, dragonflies, and butterflies (Stewart 1988).

Botany of Madre de Dios, Peru

A few floras have been completed for the Neotropics or are in progress for various areas (Frodin 2001) but these have been surprisingly few considering that the completion of local and regional floras can be the basis for many other biological studies. For example, the *Flora of Barro Colorado Island* in Panama (Croat 1978) facilitated many other investigations, some of which were summarized by Leigh (1999). The *Guide to the Plants of Central French Guiana* (Mori et al. 1997, 2002) and others such as Ribeiro et al. (1999) and Vasquez et al. (2001) have led to ecological studies and supported conservation of the study areas. Such works have promoted tourism and scientific research programs that have generated substantial support for local communities. Products like these provide significant tools and value-added components, as they facilitate the work of ecologists, conservation biologists, molecular biologists, ethnobotanists, economic botanists, zoologists, students, tourists, educators, agronomists, and others.

In his last publication from Tambopata, Gentry (1994) marveled at the "dozens of undescribed species," including one of the most common floodplain trees. Likewise, Foster (1990) wrote, "All of southern Amazonian Peru, and especially the Madre de Dios drainage, which includes parts of adjacent Brazil and Bolivia, is poorly collected." Botanical research over the past 30 years in Madre de Dios has been important for identifying the extent and significance of the region's plant diversity, producing some publications and datasets (e.g., Gentry 1982a, 1982b; Phillips & Gentry 1993a–b; Foster et al. 1994; Terborgh & Andersen 1998; Pitman et al. 1999, 2001). There have been two geographic areas of botanical investigations in Madre de Dios, with most of the resources focused on quantitative inventory of trees (trunk diameter > 10 cm) in scattered one-hectare forest plots and 0.1-hectare transects. The Cocha Cashu Biological Station in Manu National Park has been the focus of significant field research during the last three decades (Terborgh 1994), with one of the only plant checklists for the region, yet it remains unpublished. The Tambopata-Candamo Reserved Zone (Foster et al. 1994) in lower Madre de Dios near the Peru-Bolivia border has been the focus of long-term monitoring of one-hectare forest plots and complementary 0.1-hectare transect work (Phillips et al. 2003a, 2003b).

Because Peru has been worked by well-known plant collectors and has a prominently published checklist (Brako & Zarucchi 1993) its flora is often assumed to be well-studied. We argue that this is not the case. There is no central published plant list for Madre de Dios, and the Brako & Zarucchi (1993) checklist is notoriously incomplete and does not provide a list of species for each department of Peru, making the task of floristic comparisons a daunting one (van der Werff & Consiglio 2004). Moreover, botanical work in Madre de Dios to date has been biased toward floodplain and terra firme forests and has ignored botanically significant but less accessible habitats like palm swamp wetlands, grassland bogs, and forest canopies. Despite the research that has been focused on selected areas in Madre de Dios, there has been no major, long-term systematic inventory of plants and habitats. There has been no floristic guide, digital or print, to form a foundation of botanical information for the region. Furthermore, until this guide, there have been no major investigations and collections of mosses in Madre de Dios nor the Andes-Amazon region of southeastern Peru, in general.

There are roughly 1460 plant species known from the Cocha Cashu Biological Station, located in Manu National Park to the west of the Los Amigos River watershed, although the full checklist has not been published. Based on a rough estimation (Brako & Zarucchi 1993), the entire department of Madre de Dios harbors 1833 species of non-pteridophyte plants. According to Smith et al. (2005), 180 species of ferns and fern allies are known from all of Madre de Dios. Yet, from our botanical surveys *only* at the Los Amigos Biological Station we have collected at least 2500 plant species, including about 200 species of ferns and fern allies, and the 63 moss species treated in this book(http://atrium.andesamazon.org). In this sense our focus on the Los Amigos River watershed provides a new model for analyzing the state of our knowledge of Madre de Dios, in general.

The Los Amigos River Watershed

The Los Amigos River watershed in the Department of Madre de Dios, southeastern Peru, represents the pristine lowland moist forest once found throughout most of upper Amazonian South America. It encompasses 1.6 million hectares (3.95 million acres) and offers the opportunity to study rain forest as it was before the disruptive encroachment of modern human civilization. The Los Amigos watershed has suffered little deforestation, industry, agriculture, or other disruptive processes. Although the region of Madre de Dios is becoming more heavily influenced by logging, gold-mining, and colonization from the Andean highlands, the Los Amigos River watershed is still in a relatively pristine state. This is evidenced by the presence of ten primate species that live within the watershed, including the brown titi (*Callicebus brunneus*), brown capuchin (*Cebus apella*), white-fronted capuchin (*C. albifrons*), squirrel monkey (*Saimiri sciureus*), emperor tamarin (*Saguinas imperator*), saddleback tamarin (*S. fuscicollis*), spider monkey (*Ateles chamek*), monk saki (*Pithecia monachus*), red howler (*Alouatta seniculus*), and owl or night monkey (*Aotus nigriceps*).

The Los Amigos Conservation Area, Los Amigos River, Madre de Dios, Peru

Because of the importance of the Los Amigos River watershed in the Peruvian Amazon, this region is now protected by the Los Amigos Conservation Area, a private 150,000-hectare conservation concession established by the Peruvian government in 2002 in cooperation with the Amazon Conservation Association (ACA) and the Asociación para la Conservación de la Cuenca Amazónica (ACCA), two sister non-profit organizations responsible for the management of this area. The Los Amigos Conservation Area protects pristine Amazonian ecosystems ranging from vast palm swamps to floodplain and upland forests between 250–320 m elevation, and it is located between 12°49' and 12°59'S latitude and 70°07'to 70°17'W longitude. The area was strategically selected for the conservation concession project because of its diversity of flora, fauna, and habitats, and its relatively pristine condition. This work is part of the the first floristic inventory and vegetation survey of the area.

The Los Amigos Biological Station, Madre de Dios, Peru

The Los Amigos Biological Station, founded in 2001, is now one of the most active research centers in western Amazonia. It is located at the southwestern edge of the Los Amigos Conservation Concession, approximately seven kilometers up the Madre de Dios River from its confluence with the Los Amigos River. It is a state-of-the-art facility equipped with solar power, gravity-driven water pumps, dormitories, cabins, camp sites, kitchen, dining room, more than 100 km of forest trails, and laboratories for studies of flora, fauna, and ecology. The station is operated by ACCA's qualified professionals, staffed with local employees, and is fully equipped for field biological research. It is also strategically located for ease of access and entry into the Los Amigos Conservation Concession via the Los Amigos River.

OVERVIEW OF THIS BOOK

This book treats 63 species of mosses distributed among 40 genera and 28 families. The largest families are: Fissidentaceae (10 spp.), Hypnaceae (7 spp.), and Sematophyllaceae (7 spp.). The number of mosses for any restricted site of a few hectares in the Amazonian lowlands apparently does not reach more than about 50 species and is often much lower than that number (Churchill 1994).

The majority of moss collections for this guide were from the region of the confluence of the Los Amigos and Madre de Dios rivers. However, collections were also made along the Madre de Dios River between the mouth of the Los Amigos River east to Puerto Maldonado (Map 1). Our hope for this moss florula is to stimulate further floristic research in the region by providing a means to identify mosses in the field or the herbarium, thus adding to our understanding of the diversity and distribution of the mosses from this and other areas of the Amazon. Future work should contribute not only better distribution data, but may also add considerable new additions to the known moss flora of Madre de Dios. Furthermore, it is hoped that this guide will promote the use of mosses in other ways, including ecological studies and comparative floristic inventories.

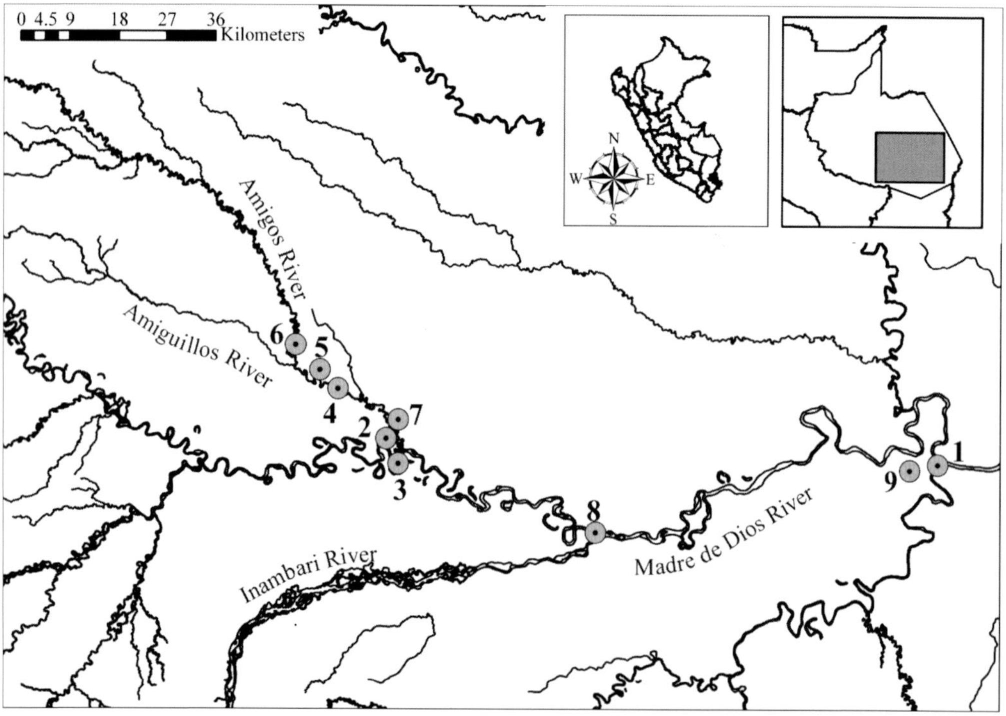

Map 1. List of collection sites and collection numbers (in *italics*) from those sites. 1. Madre de Dios: Manu Province, Puerto Maldonado, ca. 300 m, 12°36'S, 69°11'W. Coll. #'s: *3720–3727*. 2. Madre de Dios: Manu Province, Los Amigos Biological Station, ca. 2 mi N of the jct of the Madre de Dios and Los Amigos rivers, ca. 280 m, 12°34'S, 70°06'W. Coll. #'s: *3728–3902; 3951.5–4178; 4301–4364*. 3. Madre de Dios: Manu Province, Community of Boca Amigo, ca. 1.6 km N of the jct of the Los Amigos and Madre de Dios rivers on the W side of the river, ca. 220 m, 12°36'S, 70°05'W. Coll. #'s: *3903–3951*. 4. Madre de Dios District: Tambopata Province, E side of the Los Amigos River, elev. ca. 300 m, ca. 12°28'50"S, 70°11'01"W. Coll. #'s: *4179–4209*. 5. Madre de Dios District: Tambopata Province, E side of the Los Amigos River, ca. 300 m, small waterfall emptying into river, 12°27'48"S, 70°13'09"W. Coll. #'s: *4210–4215*. 6. Madre de Dios District: Tambopata Province, E side of the Los Amigos River, ca. 300 m, 12°25'37"S, 70°15'35"W. Coll. #'s: *4218–4231*. 7. Madre de Dios District: Tambopata Province, E side of the Los Amigos River, ca. 230–250 m, from banks of river to inland, 12°31'52"-12°32'05"S, 70°05'24"-70°05'29"W. Coll. #'s: *4240–4265*. 8. Madre de Dios: Manu Province, Inambari River ca. 2 km from jct of Madre de Dios River, ca. 220 m, 12°43'87"S, 69°45'51"W. Coll. #'s: *4266–4267*. 9. Madre de Dios District: Manu Province, ca. 13 km W of Puerto Maldonado in small park dominated by *Mauritea*. Collected with Euridice Honorio. Coll. #'s: *4268–4300*.

The majority of Amazonian lowland mosses are found as epiphytes, occurring on bark (corticolous or ramicolous) or leaves of trees, treelets, shrubs, and lianas (epiphyllous). Less frequently mosses are found on decomposing logs (lignicolous). Fewer still are found on rocks (saxicolous) or soil (terricolous) and are then either associated with exposed roots from fallen trees or on exposed streams and riverbanks, or with cut banks along roads or trails. There are also a few semi-aquatics or aquatics found on rocks in streams (Churchill 1994).

Collection and Identification of Mosses

Mosses can be gathered and placed in paper sacks or folded newspaper. If collections are wet they should be allowed to air dry to eliminate the possibility of fungal contamination. It is not necessary to collect plants with sporophytes for identification as most mosses can be identified from gametophytic material only. However, sporophytic material is necessary to identify several groups beyond genus, so it is always beneficial to collect sporophytic material of moss species when possible. Information about the location and date of collection, as well as the substrate that the moss is growing on should be noted on the paper collection bag or newspaper for later use in generating herbarium labels. From the collection site it is important to record elevation and geographic coordinates using a hand-held Geographic Positioning System (GPS) if possible. If sufficient material is available, collect enough to make duplicate collections that can be sent to experts for identification or to distribute to various herbaria within Peru and possibly outside of the country. The first set of all collections should be deposited in a Peruvian herbarium, preferably one closest to the study site. The second set of collections should be deposited in a national herbarium, such as the San Marcos Herbarium of Lima, Peru.

Basic equipment necessary to identify mosses includes a dissecting and a light microscope, as well as a hand lens. Dissecting needles and a pair of sharp forceps are helpful for removing leaves for viewing with the light microscope. To examine a collection, usually one or two individual plants are sufficient for most acrocarpous mosses or a portion of a stem and branch of a pleurocarpous moss. These can be placed in water—best at boiling point—and then mounted on a microscope slide in a small drop of water. Under the dissecting scope, use the dissecting needles or forceps to strip off the leaves. After this is accomplished, add a cover slip to the slide mount for viewing under the light microscope (Churchill 1994).

TAXONOMIC TREATMENT

Many of the following keys, descriptions, and illustrations have been modified or adapted from the following works: Allen 1994; Allen 2002; Buck 1998; Buck 2003; Sharp et al. 1994. Although all collections on which this manual are based [with the exception of the Vargas (1974) collections] were collected between 200–350 msl, in or near tropical, lowland, semi-deciduous, subtropical moist forest, we have provided a broader habitat description that more fully encompasses where each species might be encountered. Names which have been reduced to synonymy after the publication of the last Peruvian moss checklist (Menzel 1992) are listed after the currently accepted name.

KEY TO THE MOSSES OF MADRE DE DIOS, PERU

1. Plants distichous with leaf bases divided into vaginant lamina ____ **FISSIDENTACEAE**
1. Plants with leaves spirally arranged or sometimes complanate; leaf bases not divided.
    2. Plants whitish, laminal cells differentiated, alternating or layered between hyaline cells (leucocysts) and green cells (chlorocysts).
        3. Shoots forming a compact head of clustered short branches (capitulum); in cross section cells alternating between chlorocysts and leucocysts; plants typically of boggy or seepy sites ____ ***Sphagnum subsecundum***
        3. In cross-section lamina bi- to multistratose, chlorocysts layered between 1 or more layers of leucocysts above and below; often lignicolous or epiphytic; plants of moist to semi-dry sites ____ **LEUCOBRYACEAE**
    2. Plants not whitish or if so, laminal cells not composed of both hyaline and green cells.
        4. Leaves with large areas of basal, hyaline cells (cancellinae) ____ **CALYMPERACEAE**
        4. Leaves usually green to base except in alar regions.
            5. Costa single.

6. Plants acrocarpous.
  7. Median cells rhomboidal to broadly or narrowly hexagonal or fusiform, not incrassate/porose, smooth throughout; peristome double ________ **BRYACEAE**
  7. Median cells mostly quadrate or rounded or if rhomboidal then crassate/porose, smooth or papillose; peristome single or double.
    8. Capsules markedly long and narrow, with a narrow apophysis as long as, or much longer than urn ________ **Trematodon longicollis**
    8. Capsules various, if narrow then apophysis shorter than urn.
      9. Cells papillose.
        10. Cells unipapillose on distal cell angles; terminal stems often appearing fasciculate with several short branches; capsules subglobose; peristome double ________ **BARTRAMIACEAE**
        10. Cells multi-papillose at back on upper dorsal surface ________ **Barbula indica**
      9. Cells not papillose.
        11. Leaf margins with a distinct border of elongate cells.
          12. Border confined to the lower 1/3 to 1/2 of leaf margin ________ **Groutiella tomentosa**
          12. Border extending to apex ________ **Dolotortula mniifolia**
        11. Leaf margins not bordered with elongate cells.
          13. Leaves strongly rugose, lower cells long-rhomboidal, incrassate and porose ________ **Schlotheimia rugifolia**
          13. Leaves not rugose.
            14. Leaves ovate to ligulate ________ **Splachnobryum obtusifolium**
            14. Leaves narrowly lanceolate to lanceolate-subulate.
              15. Leaves 4–6 mm long ________ **Pilopogon guadalupensis**
              15. Leaves 1–2.5 mm long.
                16. Leaves lanceolate-subulate, margins erect to narrowly incurved; peristome teeth divided nearly to base ________ **Ditrichum difficile**
                16. Leaves broadly lanceolate, margins narrowly recurved; peristome teeth divided ca. half of their length ________ **Dicranella hilariana**
6. Plants pleurocarpous.
  17. Leaf cells papillose.
    18. Stem and branch leaves not exceeding 0.75 mm long and usually much shorter; paraphyllia usually conspicuously produced on the stem ________ **Cyrto-hypnum** spp.
    18. Leaves greater than 1 mm long; paraphyllia absent ________ **Henicodium geniculatum**
  17. Leaf cells not papillose.
    19. Leaves ligulate to spathulate ________ **Neckeropsis** spp.
    19. Leaves various but not ligulate to spathulate.
      20. Plants stipitate-frondose ________ **PTEROBRYACEAE** p.p
      20. Plants not stipitate-frondose.
        21. Stem and branch leaves not differentiated; plants terrestrial or growing on rocks or bases of trees, rarely on twigs ________ **Rhynchostegium serrulatum**
        21. Stem and branch leaves differentiated; plants growing on tree trunks and branches, rarely on tree bases and rocks.
          22. Alar cells strongly differentiated, in excavate groups ________ **Squamidium nigricans**
          22. Alar cells not or very weakly differentiated ________ **Zelometeorium patulum**
5. Costa double or absent.
  23. Alar cells differentiated, usually quadrate or inflated and oval.
    24. Leaves conduplicate ________ **Phyllogonium viscosum**
    24. Leaves not conduplicate.
      25. Alar cells unequally differentiated on one side of lateral leaves ________ **Pilosium chlorophyllum**
      25. Alar cells equally differentiated on both sides.
        26. Branch leaves not heterophyllous ________ **SEMATOPHYLLACEAE**
        26. Branch leaves heterophyllous.
          27. Alar cells numerous, quadrate, reaching the costa in narrow leaves ________ **Entodon jamesonii**
          27. Alar cells 3–5, enlarged at the extreme basal angles ________ **Rhacopilopsis trinitensis**
  23. Alar cells not differentiated.
    28. Leaves with a well developed double costa, 1/4 or more the lamina length ________ **PILOTRICHACEAE** p.p.
    28. Leaves with costa short and double, less than 1/4 the lamina length or absent.

29. Plants complanate with differentiated lateral and dorsal leaves; typically epiphyllous but sometimes on twigs, logs, or humus ______ **Crossomitrium patrisiae**
29. Plants with undifferentiated leaves or if differentiated not complanate; occurring on a wide range of substrates but generally not epiphyllous ______ **HYPNACEAE** p.p.

BARTRAMIACEAE

Plants small to robust in loose or dense tufts, mostly tomentose below, erect, with forked and fastigiate or whorled branching. **Leaves** ovate-lanceolate to linear- or lanceolate-subulate, acute or acuminate, unbordered, toothed at the margins above and often at the back of the costa; **costa** single, mostly strong, ending below the apex to excurrent as a toothed awn; cells rounded-quadrate to oblong or linear, papillose at or near 1 or both ends or sometimes centrally, rarely smooth, usually enlarged, lax, smooth and hyaline toward the base, not or only slightly differentiated at the basal angles. **Setae** terminal, short to elongate; **capsules** inclined or horizontal, subglobose or ovoid, mostly asymmetrical or oblique-mouthed, generally furrowed; opercula convex or short-conical, rarely rostrate; peristome double, single rudimentary, or lacking. **Calyptrae** cucullate, smooth, naked.

PHILONOTIS Brid.

Plants small to fairly robust, growing in wet places in loose to dense, bright-green or yellowish tufts, ± tomentose below. **Stems** irregularly branched or frequently with whorled innovations below the inflorescences. **Leaves** erect, spreading or ± secund when dry, lanceolate to ovate-lanceolate, acuminate and acute or, rarely, bluntly acute to obtuse, serrulate nearly throughout; **costa** subpercurrent to excurrent; upper cells oblong to linear, papillose at the upper or lower end (or both), very rarely smooth or centrally unipapillose. **Setae** single, straight or flexuose, elongate; **capsules** strongly inclined to horizontal, asymmetrical, subglobose, ribbed, neck short; opercula convex-conical; exostome teeth densely and finely papillose, trabeculate; endostome with a well-developed basal membrane, keeled, vertically papillose-striate segments, and rudimentary to rather well developed. **Spores** reniform, densely papillose.

1. Plants autoicous; costa long-excurrent ______ **Philonotis longiseta**
1. Plants dioicous; costa subpercurrent to short-excurrent ______ **Philonotis uncinata**

1. Philonotis longiseta (Michx.) E. Britton (Fig. 1)

Plants small to medium-sized, in loose, yellowish-green to green tufts. **Stems** red. **Leaves** 1–2.5 mm long, erect to erect-spreading when dry or erect-spreading when wet, straight or secund, lanceolate to triangular-lanceolate; apices acuminate; margins narrowly revolute, singly or doubly serrate; **costa** long excurrent; cells papillose at upper cell ends, upper cells short-rectangular, basal cells long rectangular, alar cells short rectangular or with 2–3 quadrate cells at extreme base. **Setae** 10–35 mm long, red; **capsules** 2–3 mm long, ovoid to subglobose, abruptly constricted at neck; opercula conical-mammillate. **Spores** 20–27 µm diameter.

Note.—Reported by Vargas (1974).

Habitat.—Lowland to upper montane wet forest, 300–3500 m, in bogs or trailside banks.

2. Philonotis uncinata (Schwägr.) Brid. (Fig. 2)

Philonotis crassinervia Broth. & Paris
Philonotis gracillima Ångström

Plants small to medium-sized, in loose or dense, glaucous green or yellowish-green tufts, to 40 mm high. **Stems** red. **Leaves** 0.5–2.5 mm long, imbricate, erect or erect-spreading when wet or dry, straight or secund, oblong-lanceolate to triangular-lanceolate; apices acuminate, acute or obtuse; margins narrowly revolute, bluntly or sharply serrate by two rows of teeth; **costa** subpercurrent to short-excurrent; cells papillose at upper cell ends, upper cells subquadrate, short- or long-rectangular, lax, basal cells near costa enlarged, rectangular, alar cells quadrate. **Setae** 10–30 mm long, red; **capsules** 1–2 mm long, ovoid to subglobose, opercula conical-mammillate. **Spores** 20–25 µm diameter.

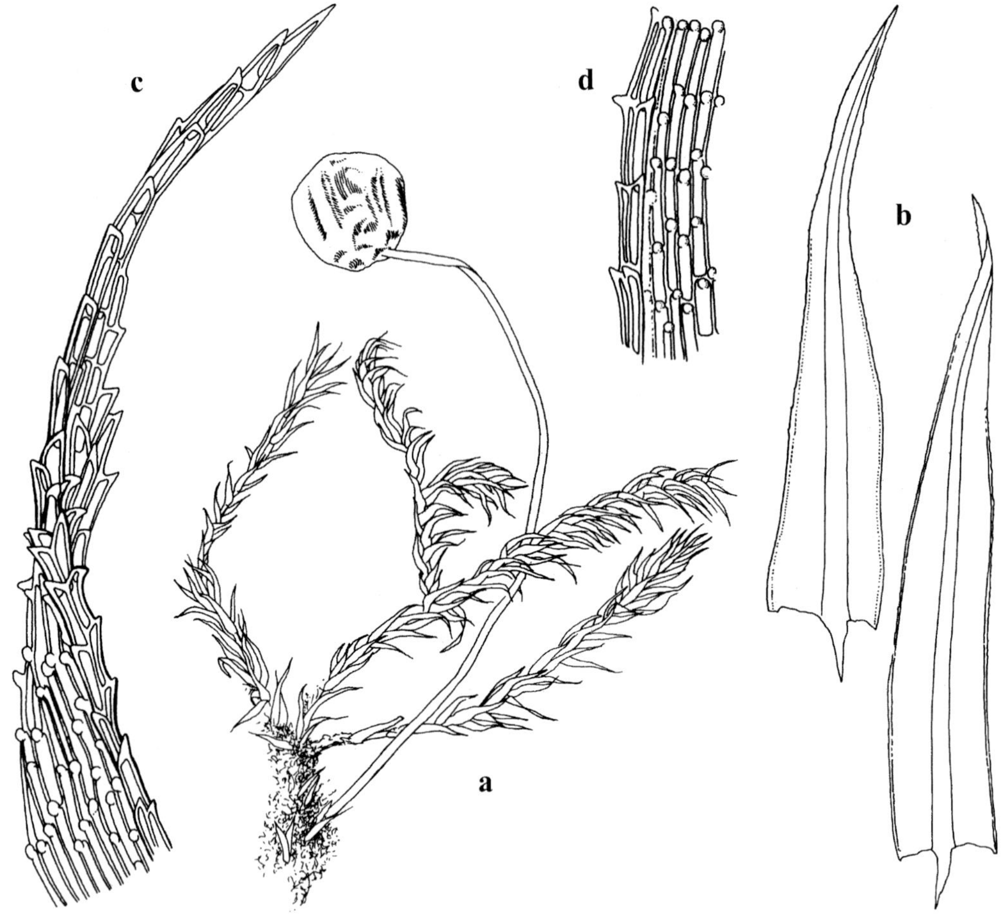

Fig. 1. ***Phylonotis longiseta*. a.** Habit, × 9. **b.** Leaves, × 70. **c.** Apical cells of leaf, × 337. **d.** Upper marginal cells of leaf, × 337. Reproduced with permission from Mem. New York Bot. Gard. 69(1): fig. 423. 1994.

Habitat.—Lowland to upper montane forest, 10–2700 m, on moist or dry clay, sandy soil, or soil banks, rocks in and along rivers, wet masonry, and on logs.

Note.—*Philonotis hastata* can also be expected to occur in the area. This species differs from the previous two in having broader (10–15 µm wide versus < 10 µm wide) upper leaf cells which are smooth to weakly papillose.

BRACHYTHECIACEAE

Plants slender to robust, in lax or dense mats or tufts, often ± shiny. **Stems** creeping to ascending, rarely erect, freely and irregularly to pinnately branched. **Leaves** appressed to erect-spreading, occasionally ± secund, mostly plicate, broadly ovate to lanceolate, ± long-acuminate (or rarely blunt or rounded at the apex); **costa** single and well developed, mostly ending below the apex, frequently projecting at the tip as a dorsal spine; cells mostly elongate, rhomboidal to linear-flexuose, smooth or, rarely, papillose at back because of projecting upper angles, laxer and often porose at the base, mostly subquadrate at the basal angles, sometimes forming well-marked alar groups. **Setae** ± elongate, frequently rough; **capsules** mostly inclined to horizontal, rarely erect and symmetrical, mostly relatively short, ovoid or oblong-cylindrical and asymmetrical, ± curved when

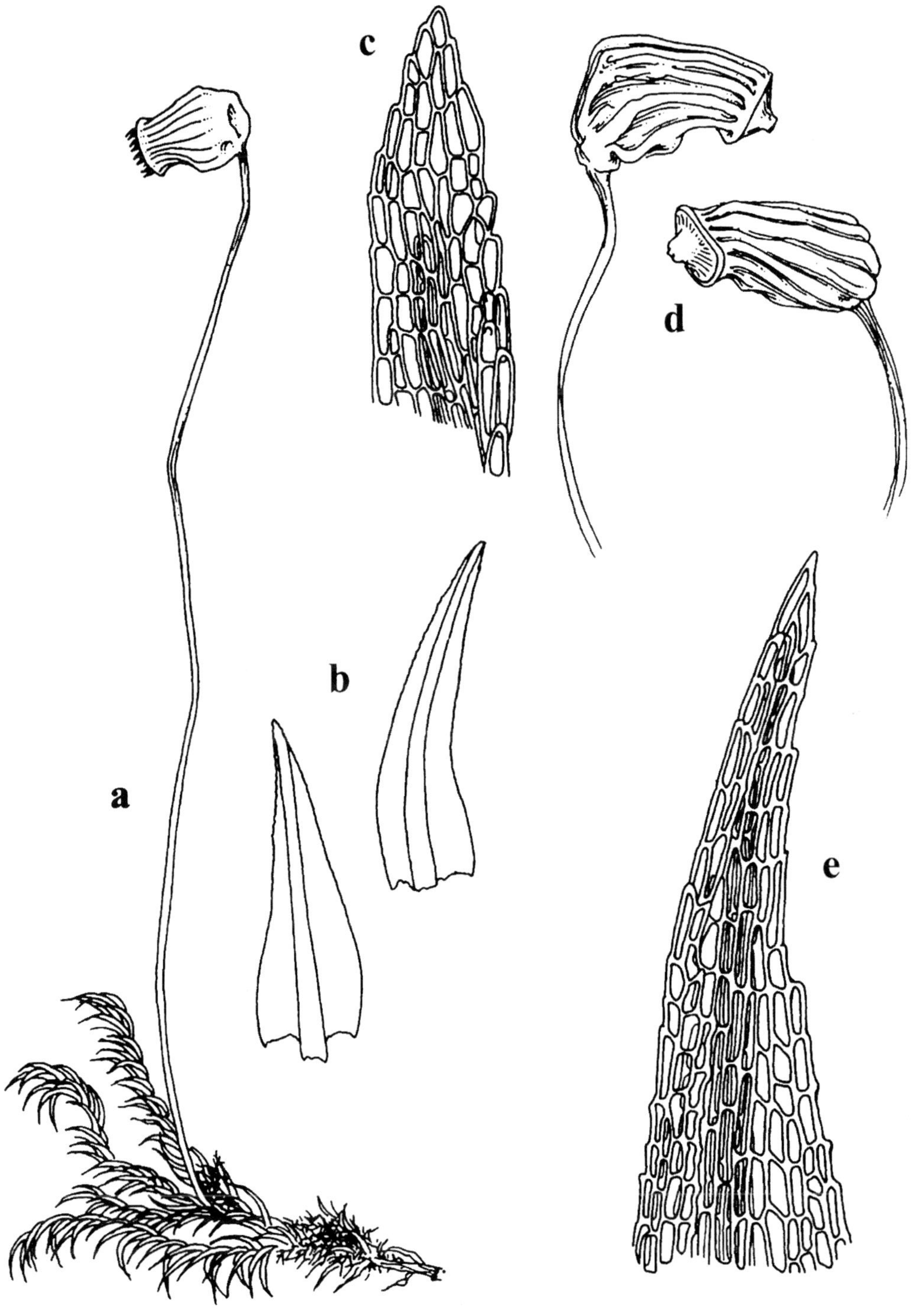

Fig. 2. *Phylonotis uncinata* var. *uncinata*. **a.** Habit. **b.** Leaves. **c.** Leaf apex. **d.** Capsules. **e.** Leaf apex of var. *glaucescens*. Reproduced with permission from Mem. New York Bot. Gard. 76(3): fig. 68. 2003.

dry and empty; annulus usually poorly differentiated; opercula conical, often rostrate; peristome double. **Calyptrae** cucullate, smooth, usually naked. **Spores** spherical.

RHYNCHOSTEGIUM Bruch & Schimp.

Plants small to rather robust, pale- to yellow-green, rarely dark-green, often in extensive mats, usually on soil. **Stems** creeping, irregularly to subpinnately branched, the branches prostrate or subascending. **Leaves** of stems and branches not or slightly differentiated, spreading to wide-spreading, rarely erect, typically ± complanate, ovate-lanceolate to ovate, usually ± abruptly acuminate, the apex often twisted, concave, not plicate; margins serrulate throughout, usually plane above, reflexed below; **costa** single, relatively slender, tapering distally, ending above midleaf, often projecting as a small spine; cells long-hexagonal to linear, smooth, thin- to firm-walled, not porose; alar cells not or gradually differentiated, then quadrate to short rectangular and lax in a broad band across the insertion. **Setae** elongate, smooth, reddish; **capsules** inclined to horizontal, ± curved, asymmetrical, cylindrical to long-cylindrical, often with a distinct neck, often constricted below the mouth when dry; operculum obliquely long-rostrate; exostome teeth shouldered, bordered, cross-striolate below on the front surface, coarsely papillose above; endostome mostly finely papillose, with a high basal membrane, segments perforate or sometimes gaping, ca. as long as the teeth. **Spores** finely papillose.

3. Rhynchostegium serrulatum (Hedw.) A. Jaeger (**Fig. 3**)

Plants of moderate size, in soft, flat, shiny, bright- or yellow-green mats, irregularly branched. **Leaves** of stems and branches wide-spreading and somewhat complanate, smooth, concave, 1.5–2.0 mm long, broadly ovate-lanceolate, slenderly acuminate, twisted at the apex; margins plane or erect, distinctly serrulate in the upper 2/3 or nearly to the base; **costa** slender, extending about 3/4 of the leaf length; cells linear-flexuose, not or weakly differentiated at the basal angles. **Setae** smooth, flexuose, yellow or orange-yellow, 10–30 mm long; **capsules** horizontal, curved-asymmetrical, oblong-cylindrical but becoming narrow, strongly contracted below the mouth, and strongly curved when dry and empty, 1.5–2 mm long; annulus well developed. **Spores** 9–16 µm diameter.

Notes.—The report of *Rhynchostegium conchophyllum* by Majestyk and Janovec (2004) is actually *R. serrulatum*.

Habitat.—In mesic to humid forest, 100–1800 m, on soil or thin soil over rocks or on humus.

BRUCHIACEAE

Plants small, annual, gregarious or loosely tufted. **Stems** erect, short, and usually simple. **Leaves** ovate or oblong-ovate and gradually or abruptly narrowed to a subula, or less commonly ovate and obtuse to broadly acute or cuspidate; margins entire or serrulate above; **costa** single, ending near the apex to short-excurrent; upper cells subquadrate to short-rectangular or oblong-linear, smooth or papillose at the back because of thickened upper ends; basal cells broader and longer, laxly oblong or oblong-rhomboid, not differentiated in the alar regions. **Setae** very short to elongate, straight or ± flexuose; **capsules** erect, broadly pyriform and inoperculate or erect to curved and cylindrical from a long, slender neck and operculate, smooth or nearly so; peristome (if present) consisting of 16 entire, forked or perforate teeth with vertical bars on the outer surface towards the base. **Calyptra** mitrate or cucullate, smooth or papillose, naked.

TREMATODON Michx.

Plants small, gregarious to loosely tufted, light-green or yellowish. **Leaves** ± contorted when dry, oblong-ovate and clasping at base, gradually to abruptly narrowed to a lanceolate or subulate point; **costa** subpercurrent to excurrent; cells small, obscure, smooth, subquadrate to short-rectangular, longer at the base. **Setae** elongate, straight or flexuose, yellow; **capsules** erect or inclined, cylindrical with a narrow neck as long as or longer than the urn, often strumose at the base; annulus simple or compound; opercula slenderly and

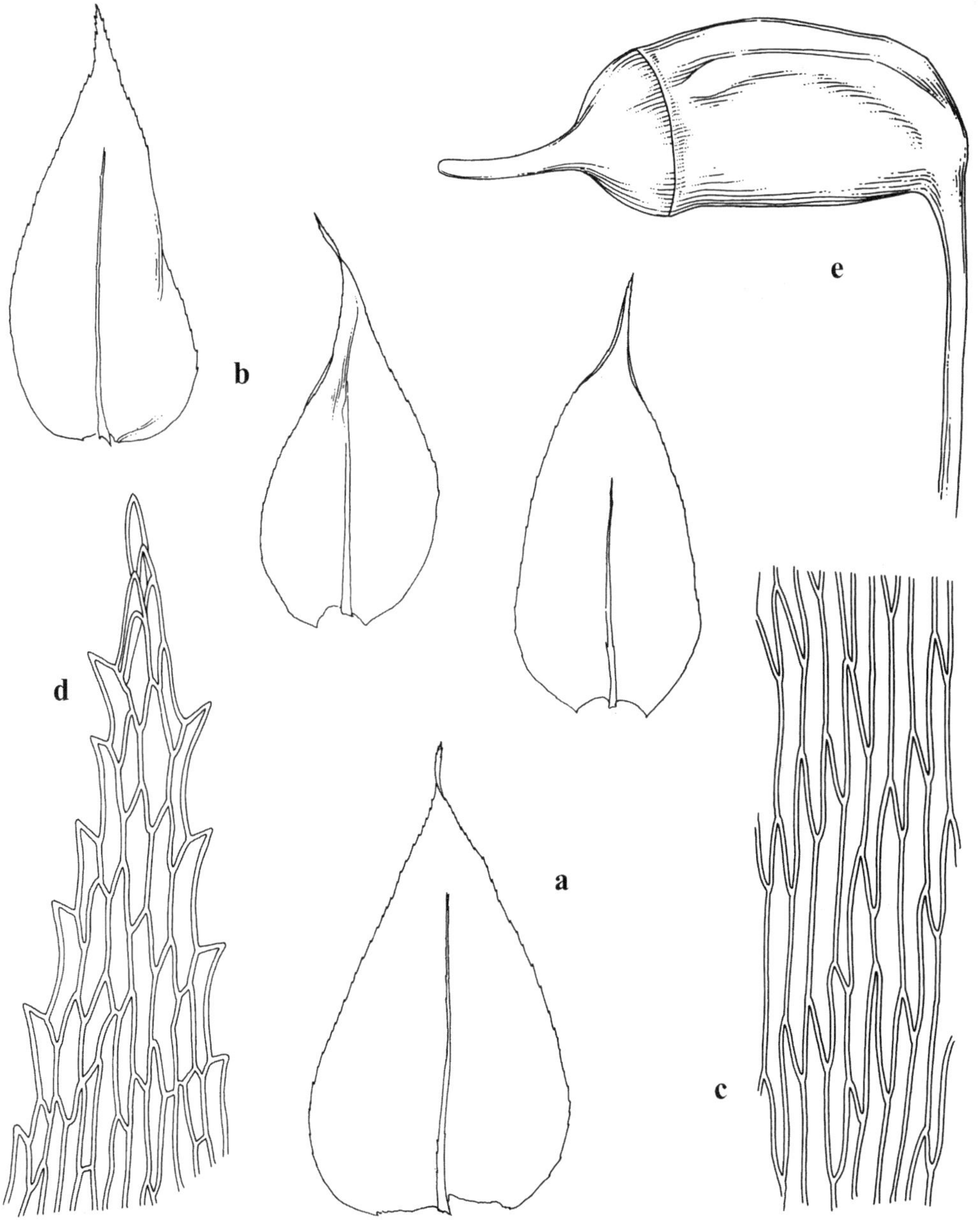

Fig. 3. *Rhynchostegium serrulatum*. **a.** Stem leaf, × 38. **b.** Branch leaves, × 38. **c.** Upper median cells of leaf, × 420. **d.** Apical cells of leaf, × 420. **e.** Capsule, × 25. Reproduced with permission from Mem. New York Bot. Gard. 69(2): fig. 699. 1994.

obliquely long-rostrate; peristome lacking or single, with 16 undivided, perforate or forked teeth. **Calyptrae** cucullate. **Spores** ± polygonal or reniform.

4. Trematodon longicollis Michx. (Fig. 4)

Plants scattered, yellow-green, terricolous, 3–5 mm long. **Leaves** evenly spaced, flexuose-spreading when wet, clasping, oblong to oblong-ovate at base, gradually long linear-subulate above, 2–3 mm long; margins weakly recurved above, entire below, weakly denticulate by blunt teeth above; **costa** percurrent; upper cells short-rectangular, walls firm and thickened; basal cells laxly oblong-rhomboidal to long-rectangular. **Setae** 10–15 mm long, smooth; **capsules** inclined, urn narrowly cylindrical, 1.5–2 mm long, neck 3–4 mm long; annuli deciduous, revoluble; peristome teeth red, lanceolate, vertically striate. **Spores** subreniform, densely papillose or minutely warty, 20–23 µm diameter.

Habitat.—Disturbed lowland to montane forest, 20–1500 m, damp sand or clay.

BRYACEAE

Plants erect, usually tufted, green, yellowish, reddish, or hyaline. **Stems** simple or forked, sometimes densely radiculose. **Leaves** elliptic to ovate or ovate-lanceolate, linear-lanceolate, lanceolate, to oblong-lanceolate, or setaceous, dimorphic and complanate or terete foliate, usually gradually increasing in size upward, often rosulate; apices usually acute, sometimes awned or piliferous; **costa** single, subpercurrent, percurrent or excurrent; margins bordered or unbordered, toothed or entire; upper leaf cells rhombic to linear or vermicular, basal cells rhombic, rectangular or quadrate. **Setae** terminal, elongate; **capsules** erect, suberect, inclined, horizontal or pendent, symmetrical or asymmetrical, cylindrical, sub-cylindrical, clavate, pyriform, oval or spherical, neck commonly differentiated; opercula convex to short-conical, rarely rostrate; annulus usually well-developed and revoluble; peristome double, usually complete, exostome of 16 triangular teeth, endostome with a basal membrane usually well-developed, segments usually broad and perforate, at times the exostome reduced or lacking. **Calyptrae** cucullate, smooth. **Spores** spherical.

1. Leaves oval to sub-orbicular, apex obtuse to rounded; costa subpercurrent; upper cells rhomboidal __________ **Bryum renauldii**
1. Leaves ovate-lanceolate to oblong or oblong-ovate, apex acute; costa subpercurrent to short-excurrent in an apiculus; upper cells linear-vermicular to linear-fusiform.
   2. Costa subpercurrent __________ **Anomobryum julaceum**
   2. Costa ending in an apiculus __________ **Bryum apiculatum**

ANOMOBRYUM Schimp.

Plants small, slender, in erect, gregarious or loose tufts. **Stems** simple or sparsely branched. **Leaves** closely imbricate, ovate or oblong, weakly decurrent; apices obtuse to rounded at the apex or broadly acute; margins plane to erect, entire to serrulate or obscurely crenulate near apex; **costa** subpercurrent to percurrent; cells smooth, thin- or thick-walled, upper cells linear-vermicular, rhomboidal or rhomboidal-hexagonal, basal cells distinctly differentiated from upper cells, broader, rectangular to quadrate, thinner-walled. **Setae** elongate, straight; **capsules** pendent, horizontal to suberect.

5. Anomobryum julaceum (Schrad. ex G. Gärtner, B. Meyer & Scherb) Schimp. (**Fig. 5**)

Anomobryum concinnatum (Spruce) Lindb.
Anomobryum filiforme (Dicks.) Husn.
Bryum concinnatum Spruce
Bryum filiforme Dicks.

Plants yellow-green, gregarious or in loose tufts. **Stems** slender, julaceus, to 25 mm high; rhizoids sparse at base. **Leaves** 1–1.2 mm long, imbricate, firm, not contorted when dry, concave, oblong to oblong-ovate; apices obtuse to broadly rounded; margins plane, not bordered, entire to weakly denticulate; **costa** subpercurrent; upper cells linear-vermicular, thick-walled, basal cells rectangular to quadrate, abruptly broader. **Setae** to 25 mm long, red; **capsules** to 1.5–3 mm long, oblong-pyriform; opercula conical-apiculate; exostome teeth yellow-brown, densely papillose, endostome hyaline, lightly papillose. **Spores** smooth, 8–20 µm diameter.

Notes.—This species commonly has a smoothly julaceus aspect, and broad, bluntly obtuse leaves with narrow, linear-vermicular, thick-walled upper leaf cells that strongly contrast with its broadly rectangular, thin-walled, often bulging, lower leaf cells. This species was reported by Vargas (1974) from Madre de Dios, but we have not seen the voucher and would not be surprised if it represented some other species.

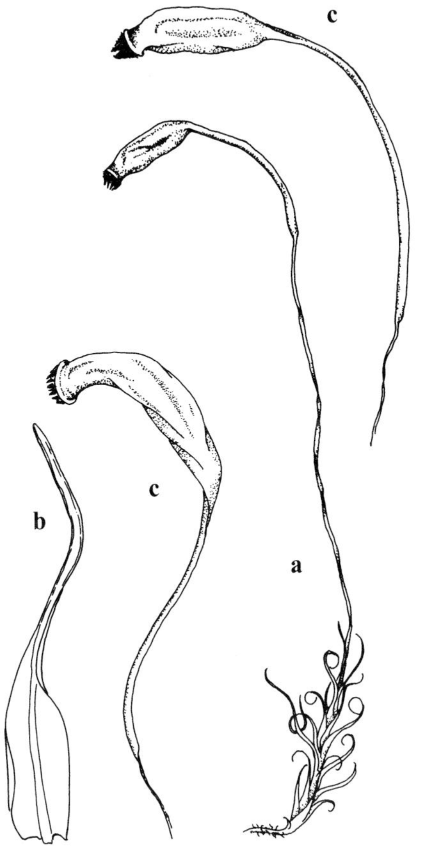

Fig. 4. ***Trematodon longicollis.*** **a. Habit, × 7. b. Leaf, × 18. c. Capsules, × 13. Reproduced with permission from Mem. New York Bot. Gard. 69(1): fig. 63. 1994.**

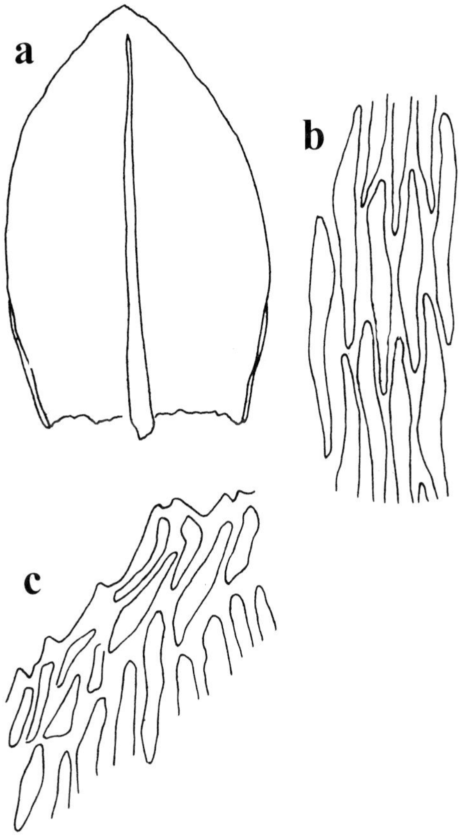

Fig. 5. ***Anomobryum julaceum.*** **a. Leaf. b. Upper cells. c. Upper marginal cells. Reproduced with permission from authors Churchill & Linares 1995.**

Habitat.—Open montane forest to páramo and puna, 1700–4570 m; exposed, dry to wet sites, on soil and soil-covered rocks.

BRYUM Hedw.

Plants small to robust, in erect, dense or loose tufts, simple or forked. **Leaves** erect to erect-spreading when dry, spreading when wet, ovate, ovate-lanceolate, lanceolate, oblong-lanceolate, oval, elliptic or suborbicular, often decurrent; apices obtuse, acute or acuminate, concave; margins plane, erect or recurved, entire to serrate near apex; **costa** strong, subpercurrent to excurrent; cells smooth, rhomboid-hexagonal to elongate rhomboidal above, quadrate to long-rectangular below. **Setae** elongate, straight or flexuose; **capsules** subcylindrical, clavate, pyriform, subglobose or ovoid, erect or nodding to pendent, symmetrical to curved; opercula conical-apiculate; annulus compound, deciduous; peristome double.

6. Bryum apiculatum Schwägr. (**Fig. 6**)

Plants small to medium-sized, in loose tufts to 3 cm high; rhizoids moderately developed at base. **Leaves** 1–2 mm long, erect to erect-spreading and well-spaced below, erect-imbricate and crowded at apex, ovate-lanceolate; apices broadly acute; margins erect below, plane above, not bordered, entire to weakly serrulate

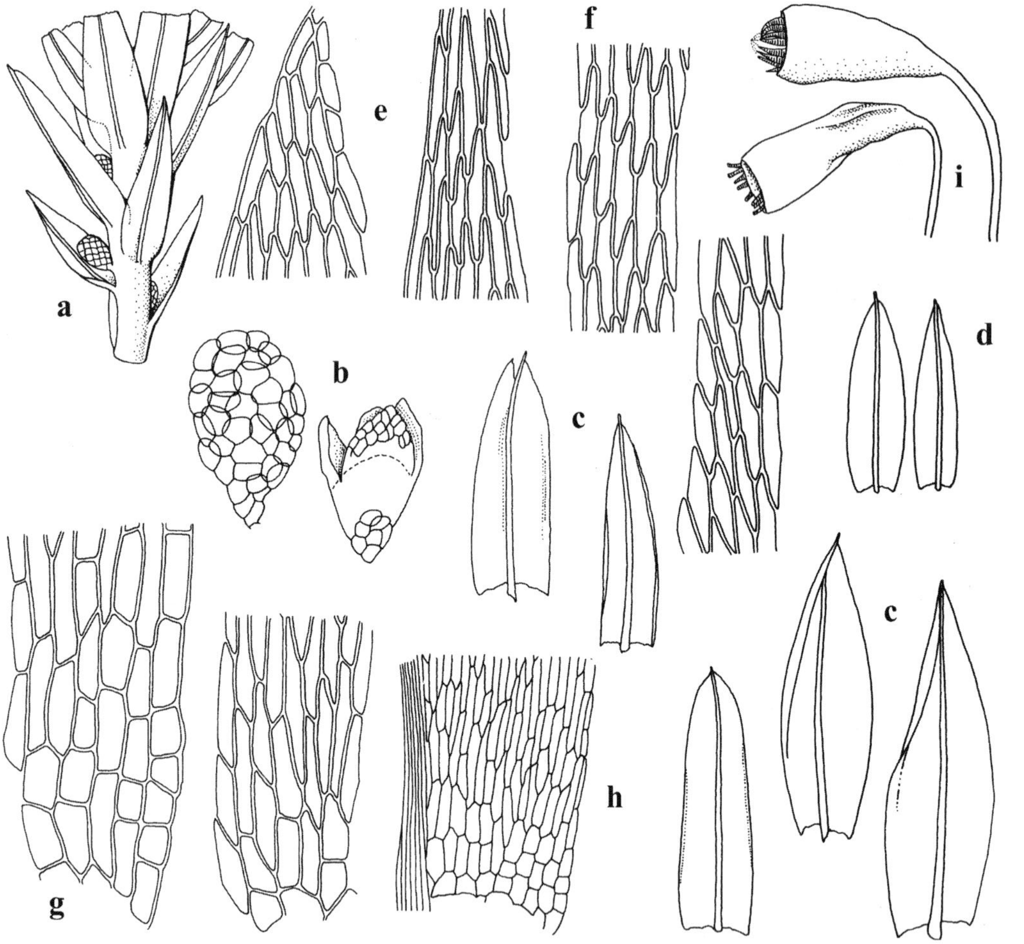

FIG. 6. *Bryum apiculatum*. **a.** Upper portion of stem with axillary brood bodies, × 14. **b.** Brood bodies, × 112. **c.** Stem leaves, × 27. **d.** Innovation leaves, × 27. **e.** Upper marginal cells of leaf, × 233. **f.** Upper median cells of leaf, × 233. **g.** Basal cells of leaf, × 233. **h.** Leaf base, × 112. **i.** Capsules, × 10. Reproduced with permission from Mem. New York Bot. Gard. 69(1): fig. 355. 1994.

above; **costa** shortly excurrent in a small apiculus; upper cells linear-fusiform, basal cells long rectangular, alar cells quadrate. Leaf axils occasionally with small brown to reddish, globose gemmae. **Setae** 15–25 mm long, red; **capsules** 1.5–2.5 mm long, red, oblong-cylindrical, inclined to pendent, opercula conical; exostome teeth orange below, yellow above, densely papillose, endostome yellowish-hyaline, lightly papillose. **Spores** lightly roughened, 12–16 µm diameter.

Notes.—Reported by Vargas (1974).

Habitat.—Wet lowland to open or wooded montane forest, 0–1700 m, on rock, brick, bark, logs, and sandy soil, along streams and waterfalls.

7. Bryum renauldii Röll ex Renauld & Cardot (**Fig. 7**)

Plants medium-sized, in loose tufts to 2.5 cm high; rhizoids sparse. **Leaves** 1.5–2.5 mm long, distant, irregularly contorted when dry, spreading when wet, oval to suborbicular, weakly decurrent; apices obtuse to rounded; margins narrowly recurved below, not or faintly bordered above, entire or obscurely serrulate;

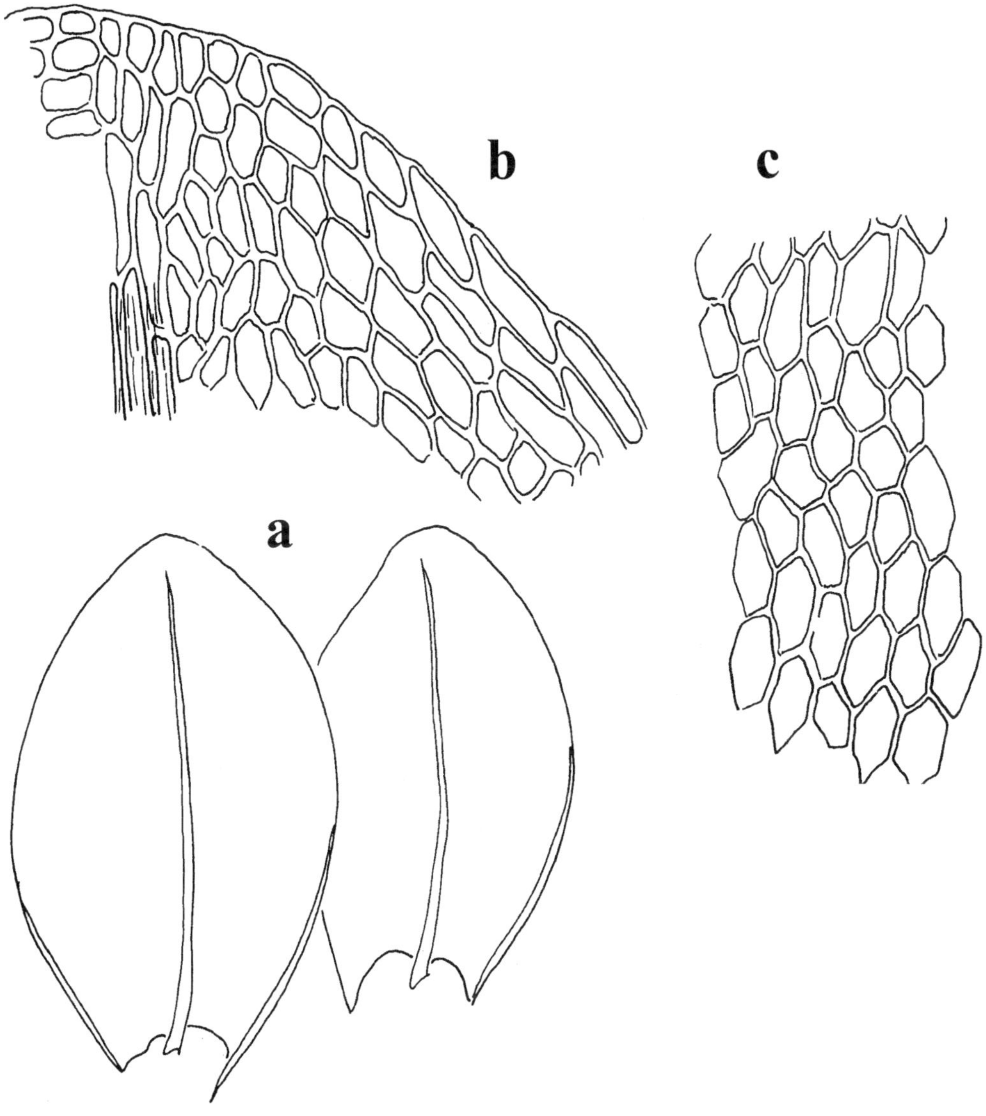

Fig. 7. *Bryum renauldii*. **a.** Leaves. **b.** Apical cells of leaf. **c.** Basal cells of leaf. Reproduced with permission from Mem. New York Bot. Gard. 69(1): fig. 356. 1994.

costa subpercurrent, ending 2–4 cells below the apex; upper cells rhomboidal, basal cells enlarged, bulging, rectangular. **Sporophytes** unknown.

Notes.—Bryum renauldii is a rheophilous species with distant, soft, obtuse, nearly entire leaves. Its upper marginal leaf cells are rectangular and somewhat longer than the irregularly quadrate inner cells, and so the leaves appear to be weakly bordered.

Habitat.—Lowlands (infrequent) to montane forest, 300–2500 m, on rocks along streams.

CALYMPERACEAE

Plants erect, rarely prostrate and with erect-ascending branches. **Stems** simple to highly forked in loose to dense mats or cushions. **Leaves** sometimes dimorphic (gemmiferous and nongemmiferous), straight to crispate or variously contorted or twisted when dry; intramarginal rows of longitudinally differentiated cells (teniolae) present or lacking; basal part mostly sheathing and with evident shoulders, including usually conspicuous areas of enlarged, hyaline, external and internally porose cells (cancellinae); upper part of leaf shorter to many times longer than lower lamina, its margins mostly variously thickened or bordered, or both, but rarely undifferentiated, mostly toothed at least in part; **costa** single and strong; cells of upper lamina isodiametric to vertically or transversely elongate, smooth, bulging or mammillose to uni- or pluri-papillose; gemmae frequently present on leaves in *Calymperes*, less common in *Syrrhopodon*. **Setae** smooth, mostly elongate; **capsules** cylindrical, mostly exerted, rarely immersed; opercula rostrate; peristome single or lacking, often vestigial, the teeth variously papillose on the front, mostly smooth internally. **Calyptrae** naked, smooth or papillose, cucullate and deciduous or rarely conical-mitrate (*Syrrhopodon*), or clasping the seta below the capsule and persistent (*Calymperes*). **Spores** mostly spherical and finely papillose.

1. Teniolae present.
   2. Upper lamina narrower than lower; cancellinae forming acute angles distally with costa _____ **Calymperes palisotii**
   2. Upper lamina often as wide as or wider than the lower lamina; cancellinae forming very broad angles distally with costa _____ **Calymperes afzelii**
1. Teniolae lacking.
   3. Upper lamina of leaves bordered in part by elongate, hyaline cells _____ **Syrrhopodon parasiticus**
   3. Upper lamina of leaves not bordered by elongate, hyaline cells.
      4. Margins of upper lamina serrate-toothed in rows, the teeth forming 3 or more irregular wings from shoulder to apex; leaf cells papillose dorsally; capsule immersed _____ **Syrrhopodon cryptocarpos**
      4. Margins of upper lamina bearing at most 2 rows of teeth; leaf cells mostly smooth dorsally (sometimes somewhat papillose); capsules exerted on a long seta _____ **Syrrhopodon incompletus**

CALYMPERES Sw. in Web.

Plants mostly tufted. **Stems** erect, mostly simple. **Leaf** margins generally thickened but mostly lacking elongate, hyaline cells; cells of upper lamina isodiametric or transversely elongate; teniolae, gemma receptacles and highly modified gemmiferous leaves common. **Capsules** exerted; peristomes lacking. **Calyptrae** plicate, persistent, enclosing the capsule, clasping the opercula by its rostrum and twisting around the seta below, bearing vertical slits above through which the spores escape.

8. Calymperes afzelii Sw. (Figs. 8, 62)

Plants yellowish to dark-green or brown, dull, in low, thin turfs or gregarious and tufted, mostly to ca. 10 mm tall. **Leaves** 3–6 mm long, only lightly contorted when dry, ligulate to lanceolate from broader base; margins thickened, serrate to coarsely toothed above; teniolae very conspicuous; cancellinae distinct; cells at midleaf obscure, smooth or finely papillose dorsally; bulging papillose ventrally, 4–5 µm in diameter; **costa** smooth or papillose; gemmiferous leaves often abruptly constricted at apex and forming a narrow "proboscis" with margins revolute around costa, gemmae borne only on ventral surface of costa. **Setae** 4–5 mm long; **capsules** 2–2.5 mm long; opercula 0.75 mm long. **Calyptrae** 3.5–4 mm long. **Spores** finely papillose, 19–23 µm diameter.

Habitat.—Coastal and interior forests, 10–750 m, on tree trunks, logs, rock and soil.

9. Calymperes palisotii Schwägr. (Fig. 9)

Plants dark- to yellowish-green, to 10 mm tall but commonly shorter, in low, dense turfs or clumps. **Leaves** dimorphic, involute and contorted when dry, mostly 2–3 mm long, oblong to broadly lanceolate, base not or only somewhat broader than upper lamina (except in gemmiferous leaves); margins usually slightly thickened, entire above, more or less serrate at shoulders; teniolae usually conspicuous at shoulders and above but sometimes weak or nearly lacking in some leaves; cancellinae distinct, often truncate distally; cells at midleaf mostly 7–8 µm in diameter, smooth or slightly papillose dorsally, bulging ventrally; **costa** smooth except towards apex; gemmiferous leaves narrowed distally, often terminating in a distinct gemmae-receptacle

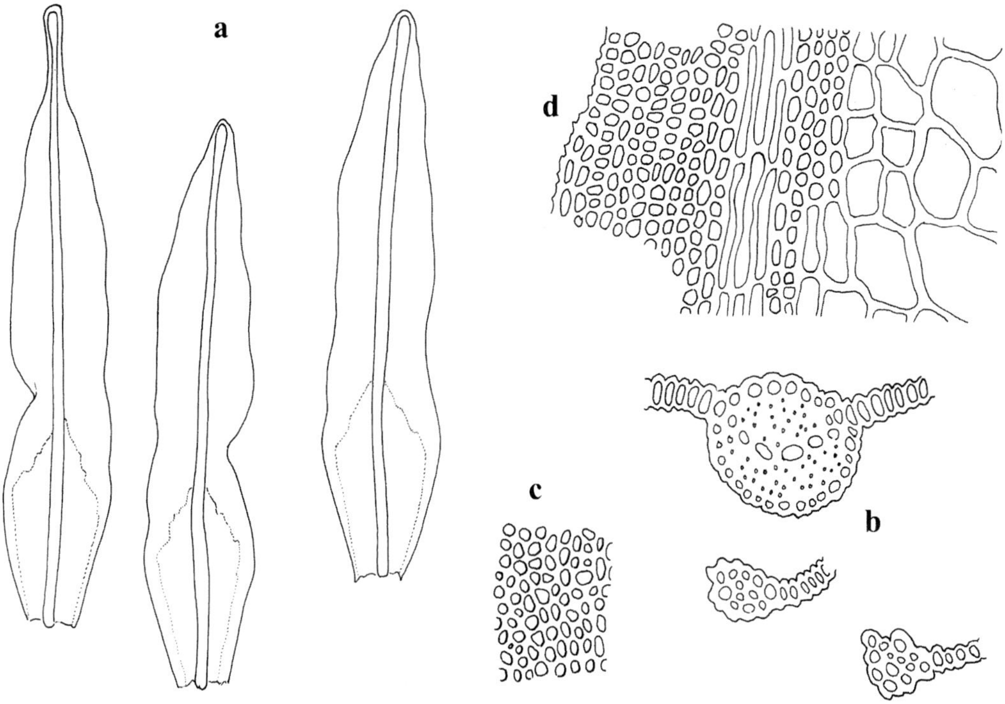

FIG. 8. *Calymperes afzelii.* **a.** Leaves, × 23. **b.** Costa and leaf margins in section, × 370. **c.** Upper cells of leaf, × 370. **d.** Cells of leaf shoulders, × 370. Reproduced with permission from Mem. New York Bot. Gard. 69(1): fig. 156. 1994.

bearing gemmae in hairbrush-like tufts on ventral surface. **Setae** ca. 3 mm long; **capsules** to 2 mm long; opercula ca. 0.5 mm long. **Calyptrae** 3–3.5 mm long. **Spores** finely papillose mostly 28–35 µm diameter.

Habitat.—Coastal and inland forest, 10–1800 m, on tree trunks, logs, and rocks.

SYRRHOPODON Schwägr.

Plants mostly tufted or gregarious on tree trunks, logs, rock, and humus. **Stems** erect or rarely repent, simple or variously branched. **Leaf** cells mostly isodiametric, rarely transversely or vertically elongate; leaf margins variously toothed, mostly thickened, bordered entirely or in part by elongate hyaline cells, or unbordered; teniolae and gemmae receptacles mostly lacking; gemmiferous leaves generally similar to vegetative leaves. **Capsules** mostly exerted, rarely immersed; peristomes mostly imperfectly developed, sometimes lacking. **Calyptrae** cucullate or rarely mitrate-campanulate, deciduous.

10. Syrrhopodon cryptocarpos Dozy & Molk. (Fig. 10)

Plants small, slender, dull green to yellowish-green, in low, thin, turfs. **Stems** to 2.5 cm tall but typically much shorter; rhizoids purple. **Leaves** mostly 2.5–3.5 mm long, somewhat contorted and often secund at stem tips when dry, ligulate to elliptic or oblanceolate from scarcely broader base, apex blunt; margins thickened, coarsely serrate-toothed in rows forming three or more irregular wings; cancellinae small, narrow, generally persistent, reaching to or beyond leaf shoulders, cells of cancellinae often more or less eroded. Median leaf cells circular to rectangular, 9–14 × 5.5–8.5 µm, sharply papillose dorsally, mammillose to mammillose-papillose ventrally; gemmae infrequently produced, in low, dense clusters along ventral surface of costa and occasionally on thickened leaf margins near the apex. **Setae** ca. 1 mm long; **sporophyte** immersed; completely enclosed by the vaginula; **capsules** urceolate, to 1 mm long; opercula rostrate to conical-rostrate,

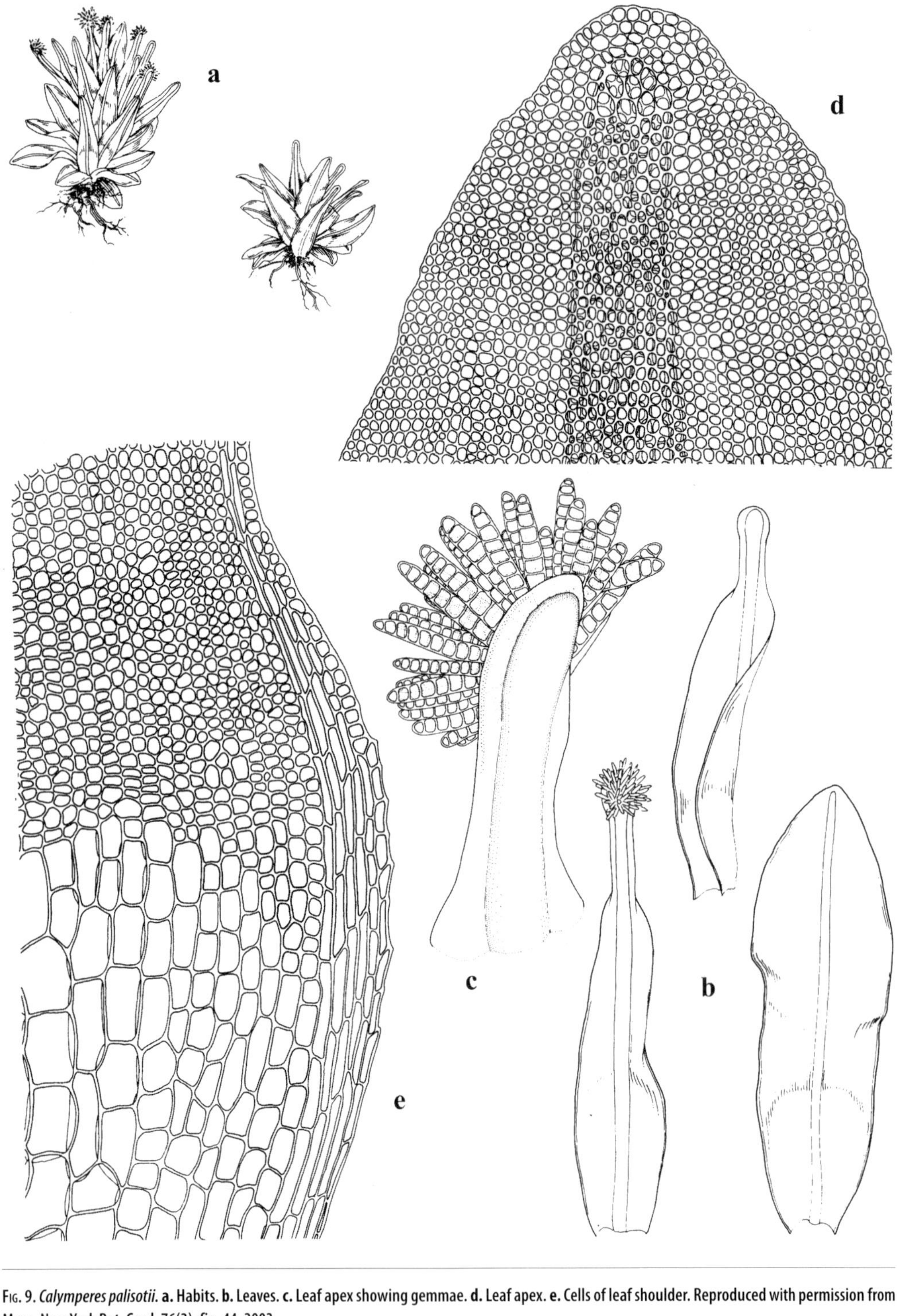

FIG. 9. *Calymperes palisotii.* **a.** Habits. **b.** Leaves. **c.** Leaf apex showing gemmae. **d.** Leaf apex. **e.** Cells of leaf shoulder. Reproduced with permission from Mem. New York Bot. Gard. 76(3): fig. 44. 2003.

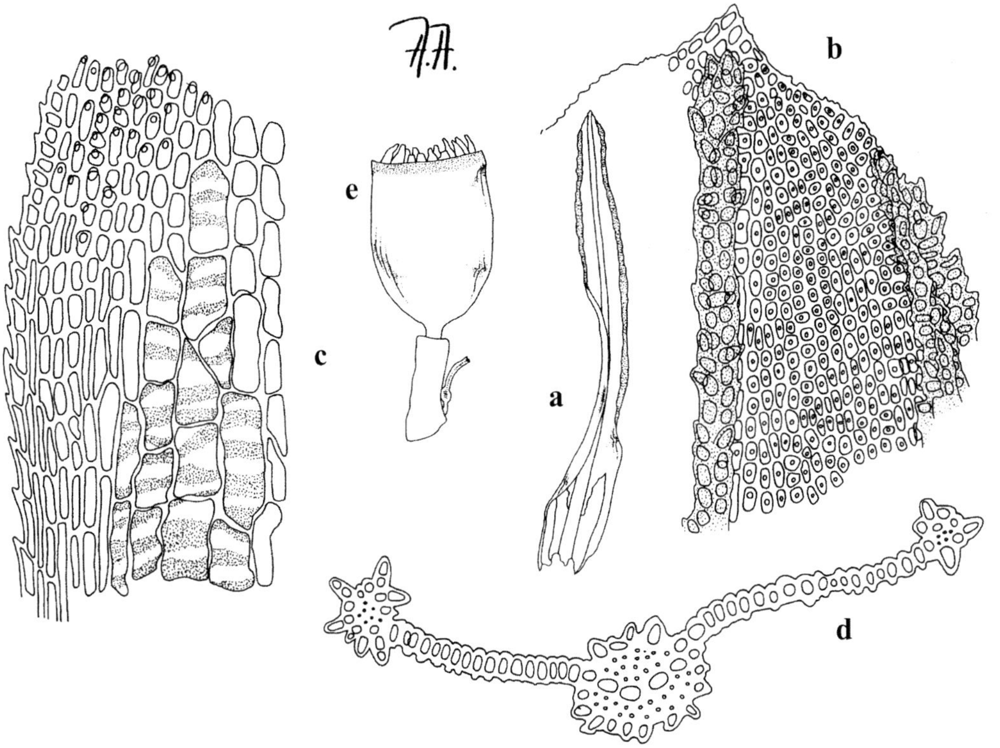

FIG. 10. *Syrrhopodon cryptocarpos*. **a.** Leaf. **b.** Leaf apex. **c.** Cells of leaf base. **d.** Leaf cross-section. **e.** Sporophyte. Drawn by Bruce Allen. Reproduced with permission from Mem. New York Bot. Gard. 76(3): fig. 49. 2003.

peristome fragile, teeth irregular. **Calyptrae** campanulate-conical, papillose distally. **Spores** smooth or granular, 19–24 µm diameter.

Habitat.—Lowland wet forest, 0–700 m, on tree trunks, logs, stumps, and termite nests.

11. Syrrhopodon incompletus Schwägr. (Fig. 11)

Plants dark-green to brownish, tufted or gregarious. **Stems** to 2–3 cm tall but often much shorter, often freely forked; rhizoids brown to purple. **Leaves** basically monomorphic, lanceolate to linear-lanceolate or nearly subulate from a broad, clasping base, 4–8 mm long, variously curved-contorted to straight and appressed when dry; margins thickened, with mostly paired teeth above; leaf shoulders often with intramarginal teniola-like features; cancellinae persistent, hyaline (rarely yellowish); cells of upper lamina obscure, quadrate to elliptical, mostly 5–6 × 5–12 µm, smooth or somewhat papillose dorsally, bulging ventrally; gemmae borne ventrally on costa at leaf tips. **Setae** yellowish-brown, 7–12 mm long; **capsules** 1.5–2.5 mm long; opercula slenderly long-rostrate; peristome lacking or rudimentary and of fragile, eroded teeth of only a few segments. **Calyptrae** 2.5–3.5 mm long, smooth. **Spores** nearly smooth to finely roughened, 14–19 µm diameter.

Habitat.—Humid lowland to lower montane forest, 0–1700 m, on tree trunks, logs, and humus; especially common on palm trunks. This is a very common moss throughout lowland tropical America.

12. Syrrhopodon parasiticus (Brid.) Besch. (Fig. 12)

Plants dark-green, solitary, in clusters of a few or gregarious. **Stems** to 2 cm tall but typically shorter, not much forked; rhizoids brown. **Leaves** to 5 mm or more long, involute and incurved when dry, basically

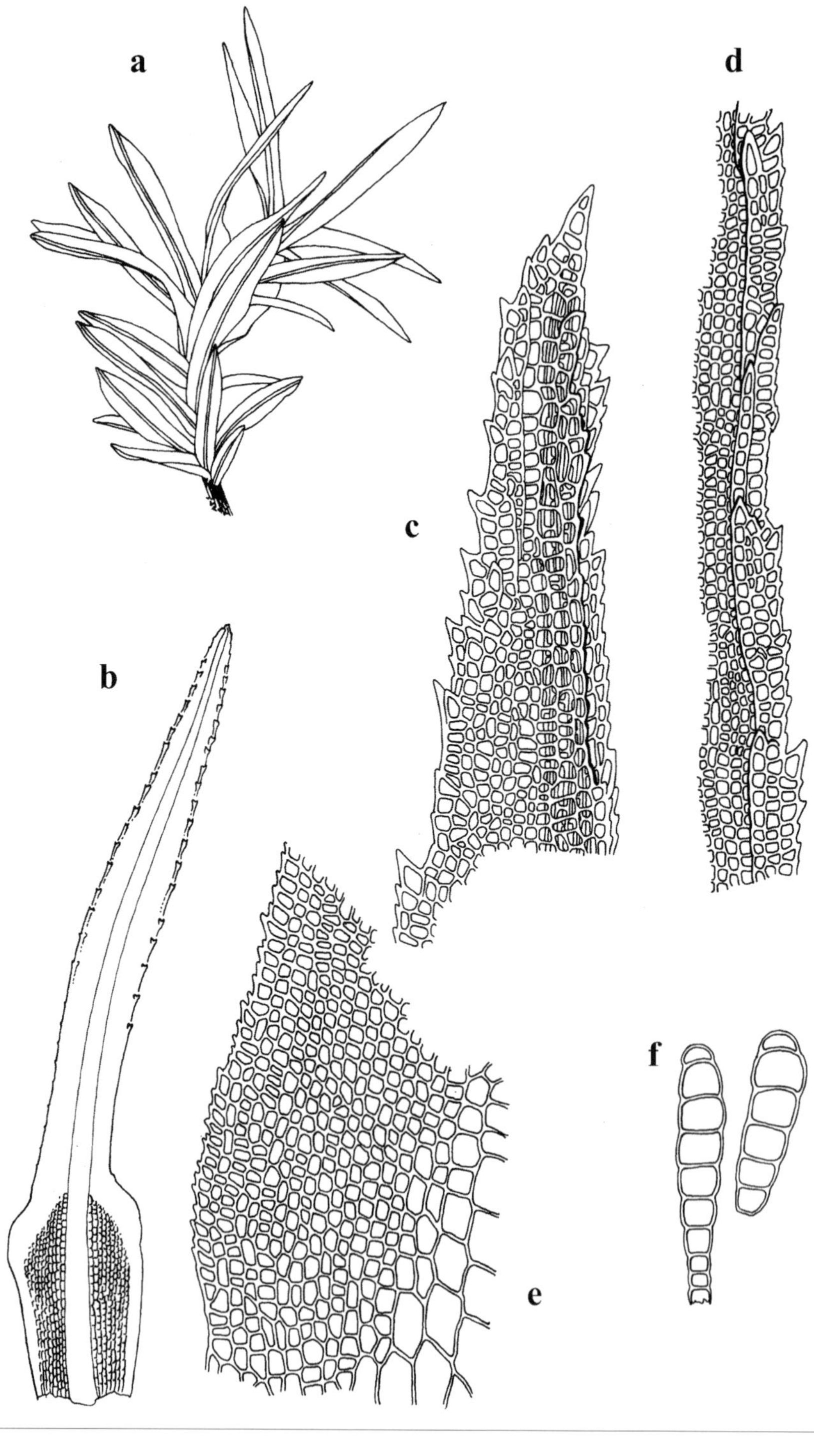

Fig. 11. *Syrrhopodon incompletus*. Var. *incompletus*. **a.** Habit. **b.** Leaf. **c.** Leaf apex. **d.** Upper marginal laminal cells. **e.** Cells of leaf shoulder. **f.** Gemmae. Reproduced with permission from Mem. New York Bot. Gard. 76(3): fig. 55. 2003.

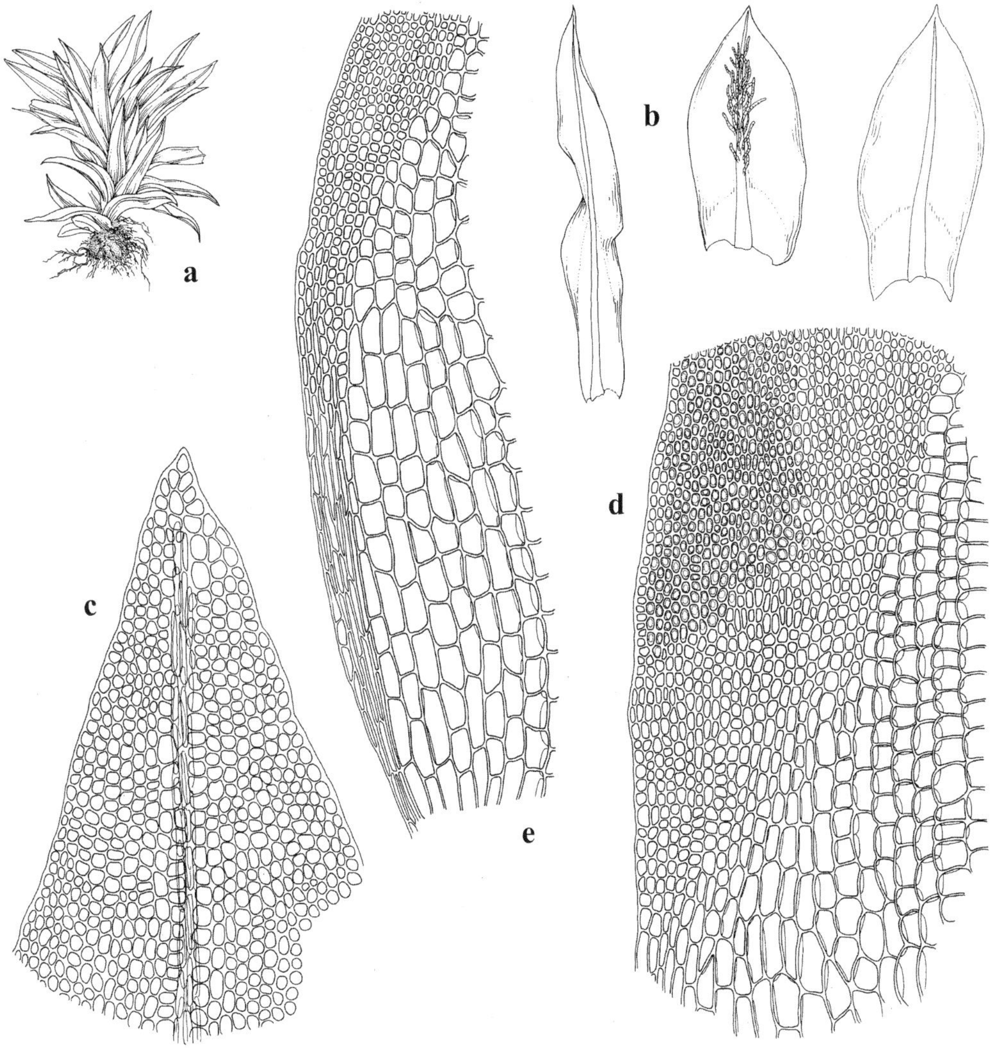

Fig. 12. *Syrrhopodon parasiticus*. **a.** Habit, × 9. **b.** Leaves, × 33. **c.** Apical cells of leaf, × 289. **d.** Cells at leaf shoulder, × 289. **e.** Cells at leaf base, × 289. Reproduced with permission from Mem. New York Bot. Gard. 69(1): fig. 149. 1994.

dimorphic, vegetative ones lingulate-lanceolate, gemmiferous ones broadly lanceolate to triangular and clustered in comae at stem tips; margins entire, mostly bordered at least in part with elongate, hyaline cells, sometimes unbordered; cancellinae distinct, distally acute; median cells ca. 7 µm, smooth to unipapillose dorsally and ventrally; gemmae filamentous, borne ventrally along costa from near cancellinae to leaf tip. **Setae** reddish, 3–4 mm long; **capsules** 1–2 mm long; peristome teeth irregular, blunt. **Calyptrae** 3 mm long. **Spores** 24–36 µm diameter, irregular in shape, granular.

Notes.—This species is easy to recognize by its very acute cancellinae, dimorphic leaves, often incomplete border, and filamentous gemmae.

Habitat.—Humid lowland to lower montane forest, 0–1600 m, strict epiphyte on twigs, branches, trunks, and leaves.

DICRANACEAE

Plants small to robust, in loose to dense, usually radiculose tufts. **Stems** often erect, simple or forked. **Leaves** in many rows, falcate-secund or not, mostly narrowly oblong-lanceolate, often broader at the base, subulate in the upper half; margins plane or recurved, entire or serrulate to serrate; **costa** single, usually strong and long, ending near the leaf apex or excurrent. Upper leaf cells variable in shape, smooth, rarely mammillose or papillose, walls entire or porose; lower laminal cells often elongate, alar cells often clearly differentiated. **Setae** mostly elongate, rarely very short, straight, flexuose, or cygneous; **capsules** usually exerted, erect, curved or inclined, ovoid, cylindrical or pyriform, sometimes strumose at base; opercula generally differentiated, obliquely rostrate from a conical base; annuli present or absent, simple to compound; peristome single, teeth rarely lacking, usually present, well-developed, 16, often reddish, deeply split into 2(–3) segments, sometimes entire. **Calyptra** cucullate.

DICRANELLA (Müll. Hal.) Schimp.

Stems simple or often forked by innovations. **Leaves** usually tapering to long-setaceous points, margin serrulate above or entire except near apex; costa occupying a large part of the subula, occasionally with several teeth at back, above. Median cells of sheathing base rectangular to linear, larger towards leaf base; not all cells differentiated. **Setae** elongate, smooth; **capsules** erect or inclined, oblong or cylindrical-oblong, usually with small apophysis, occasionally strumose; opercula conical-rostrate; annulus large; peristome teeth 16, linear-lanceolate, reddish-brown below, undivided or divided to below the middle into 2 papillose forks.

13. Dicranella hilariana (Mont.) Mitt. (**Fig. 13**)

Plants small, dull green to yellowish green; **stems** 3–25 mm tall. **Leaves** erect-flexuose to subsecund, narrowly lanceolate, gradually tapered to a broadly rounded or obtuse apex; margins narrowly recurved, entire below, serrulate at apex; costa percurrent; upper leaf cells subquadrate to short rectangular; lower cells rectangular. **Setae** yellowish, becoming red with age, straight or flexuose, 5–16 mm long; **capsules** erect, ovoid to subcylindric; annuli compound, revoluble; peristome 0.25–0.35 mm long, striate-papillose below, papillose and hyaline above, divided more than halfway to the base. **Spores** papillose, 16–20 µm diameter.

Notes.—This species is characterized by the upper leaves that are broadly lanceolate and triangular at base with obtuse or broadly acute apices and narrowly recurved margins. Reported by Vargas (1974).

Habitat.—Humid or wet lowland to lower montane forest, 0–2500 m, usually exposed sites, rather common along banks of rivers, roads, streams, and trails, and on clay, rocks, and sandy soil.

PILOPOGON Brid.

Stems mostly conspicuously equally foliate when dry, giving the plants a filiform appearance, erect spreading when moist, sparsely radiculose at base. **Leaves** straight, 3–7 mm long, lanceolate, long subulate. **Costa** at base 1/5 to 1/2 the leaf width, excurrent in a short awn, smooth, denticulate at the extreme tip, or excurrent in a long serrate hyaline awn. Lower laminal cells thin-walled, hyaline, rectangular, narrower at margins, rarely incrassate, upper laminal cells thick-walled, oblique, rhomboidal, oval, elongate-oval, or short rectangular; alar cells not differentiated, rarely inflated and reddish. **Setae** erect; **capsules** cylindrical; opercula long rostrate, erect; annulus present, dehiscent; peristome teeth 16, on a basal membrane, undivided or split to the base, densely papillose. **Spores** smooth to minutely papillose.

14. Pilopogon guadalupensis (Brid.) J.-P. Frahm (**Fig. 14**)

Plants variable in size, from 0.5 to 15 cm tall, generally 2–5 cm, yellowish to dirty or brownish green. **Stems** very slender, equally foliate with appressed leaves. **Leaves** narrowly lanceolate, 3–5 mm long, with excurrent costa, smooth at margins. **Costa** occupying 1/2 of the leaf base, ribbed at back, excurrent in a serrate awn. Alar cells indistinct or absent. Lower laminal cells 40–60 × 6–9 µm, upper laminal cells elongate oval, 20–28 × 4–8 µm. **Setae** 10–20 mm long, nearly totally surrounded by the perichaetial leaves; **capsule** erect

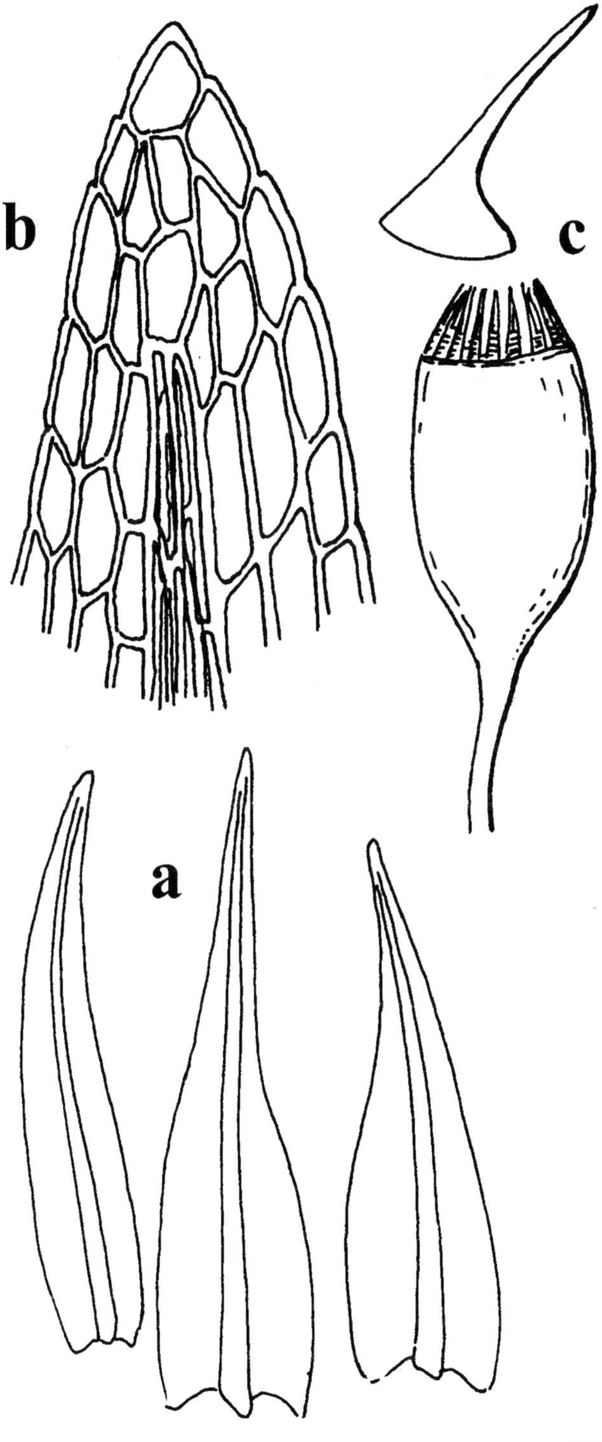

FIG. 13. *Dicranella hilariana*. **a.** Leaves, × 29. **b.** Apical cells of leaf, × 481. **c.** Capsule, × 23. Reproduced with permission from Mem. New York Bot. Gard. 69(1): fig. 82. 1994.

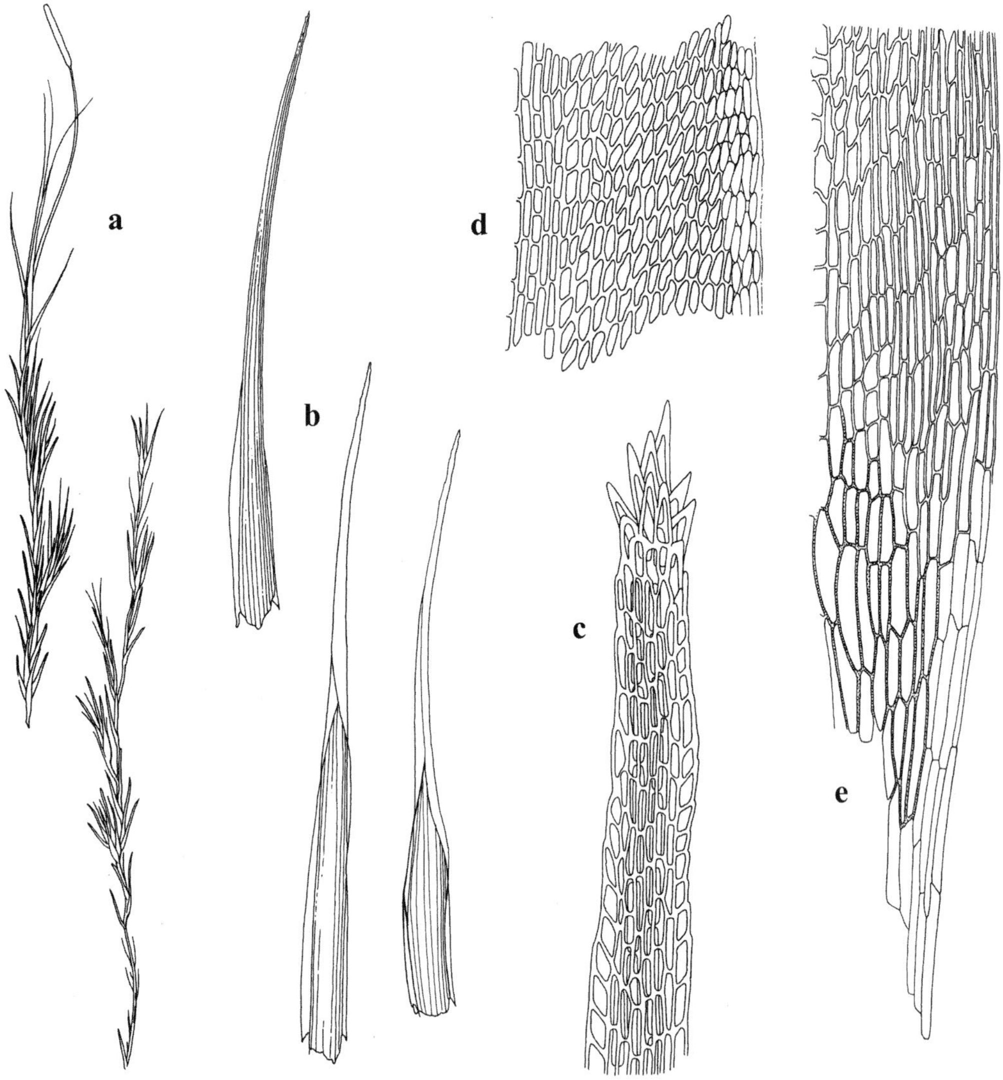

Fig. 14. ***Pilopogon guadalupensis*. a. Habits, male and female plants, × 2. b. Leaves, × 27. c. Apical cells of leaf, × 309. d. Upper marginal cells of leaf, × 309. e. Basal cells of leaf, × 200. Reproduced with permission from Mem. New York Bot. Gard. 69(1): fig. 103. 1994.**

or sometimes ± asymmetrical and curved, 2–4 mm long; peristome teeth undivided, filiform. **Calyptrae** ciliate at base. **Spores** smooth, 10–13 μm diameter.

Notes.—Reported by Vargas (1974). Very unlikely to occur in the lowland areas.

Habitat.—Open montane to páramo, 1100–4400 m, on open bare rocky soil along trails and road banks.

DITRICHACEAE

Plants mostly minute to small, or at times to medium-sized, gregarious or in loose tufts, sometimes densely caespitose. **Stems** erect, simple or branched. **Leaves** usually lanceolate, acuminate or subulate, sometimes sheathing at base, seldom distichous; margins plane, reflexed or revolute, entire or serrulate; **costa** single, narrow, subpercurrent to long-excurrent; cells sub-quadrate or short-rectangular to linear, mostly smooth or occasionally mammilose; alar cells not differentiated. **Sporophytes** terminal; **setae** usually elongate, straight, sometimes very short; **capsules** often long-exerted, at times immersed to emergent, subglobose to cylindrical, erect to inclined or curved, asymmetrical; operculum often differentiated, conical or long-rostrate, sometimes absent; annuli mostly differentitated; peristome lacking or present with 16 teeth, divided in terete divisions or perforated nearly to base, without vertical pitted-striolation or with oblique striolation. **Calyptrae** cucullate, rarely mitrate. **Spores** small, spherical, minutely papillose.

DITRICHUM Hampe

Plants small to medium-sized. **Stems** mostly simple or nearly so. **Leaves** appressed or homomallous, tapering to a spreading, subulate point from an erect or sheathing base, the lamina narrow above; margin erect or convolute, almost entire, occasionally with a few teeth at the apex; **costa** broad, occupying almost the entire width of the upper lamina, percurrent to excurrent, smooth or papillose at the back. Median laminal cells narrowly rectangular; cells of sheathing base rectangular, thin- or thick-walled. **Setae** elongate, straight, smooth; **capsules** erect or ± inclined, mostly oblong-cylindrical, symmetrical or ± curved with a poorly defined apophysis, furrowed when dry; opercula conical or rostrate, blunt at apex; annulus large; peristome teeth 16, each tooth deeply split into two filiform segments, reddish-brown, densely papillose. **Calyptrae** cucullate.

15. Ditrichum difficile (Duby) M. Fleisch. (**Fig. 15**)

Plants slender, silky, densely tufted, yellowish-green; stems 1–2 cm high, sparsely to densely tomentose. **Leaves** erect, flexuose, 2–3 mm long, lanceolate-subulate, from a narrow subclasping base; margins erect to narrowly incurved, entire or faintly serrulate; **costa** percurrent to short excurrent with a few blunt teeth at the apex; cells firm-walled to incrassate, basal cells linear, 30–60 µm long, upper cells short-rectangular, 10–25 µm long. **Setae** 10–20 mm long, yellow, becoming reddish with age; **capsules** erect to inclined, cylindrical, 2 mm long; opercula conical-rostrate, 1 mm long, red; peristome teeth 0.5–0.6 mm long. **Spores** nearly smooth, 15 µm diameter.

Habitat.—Lowland to high montane forest and subpáramo, 300–4150 m, in shaded to more commonly exposed sites, on soil and rocks, often found on steep banks.

ENTODONTACEAE

Plants medium-sized, rarely robust, in thin to dense, flat mats. **Stems** creeping, freely to sparsely but irregularly branched, terete- or complanate-foliate. **Stem and branch leaves** similar, crowded, lanceolate to broadly ovate, obtuse, acute or acuminate, usually concave, rarely plicate; margins usually serrulate above, ± entire below, mostly plane, rarely recurved; **costa** double, to midleaf or usually shorter, or absent; cells linear, smooth, often becoming shorter in the extreme apex; alar cells numerous, subquadrate to oblate. **Setae** elongate, smooth, red or yellow; **capsules** erect, symmetrical, cylindrical; opercula conical to conical-rostrate; peristome double. **Calyptrae** cucullate, naked or sparsely hairy, smooth. **Spores** spherical, smooth to coarsely papillose.

ENTODON Müll. Hal.

Plants medium-sized to moderately robust, in mostly soft, thin to dense, flat, often extensive mats. **Stems** creeping or occasionally ascending, irregularly to subpinnately branched, with branches relatively short, terete- or complanate-foliate. **Leaves** oblong-lanceolate to oblong-ovate, obtuse to acuminate, usually acute, concave, not plicate, not decurrent; margins usually serrulate above, entire below or rarely throughout; **costa** double, very short to ending near midleaf; cells linear, straight to ± flexuose, smooth, often becoming shorter at the extreme apex; alar cells numerous, quadrate to subquadrate. **Setae** red or yellow; annulus differentiated or not; exostome teeth yellow-brown to reddish, variously striate or papillose, often perforate; endostome with a low basal membrane, mostly perforate, rarely rudimentary. **Spores** papillose.

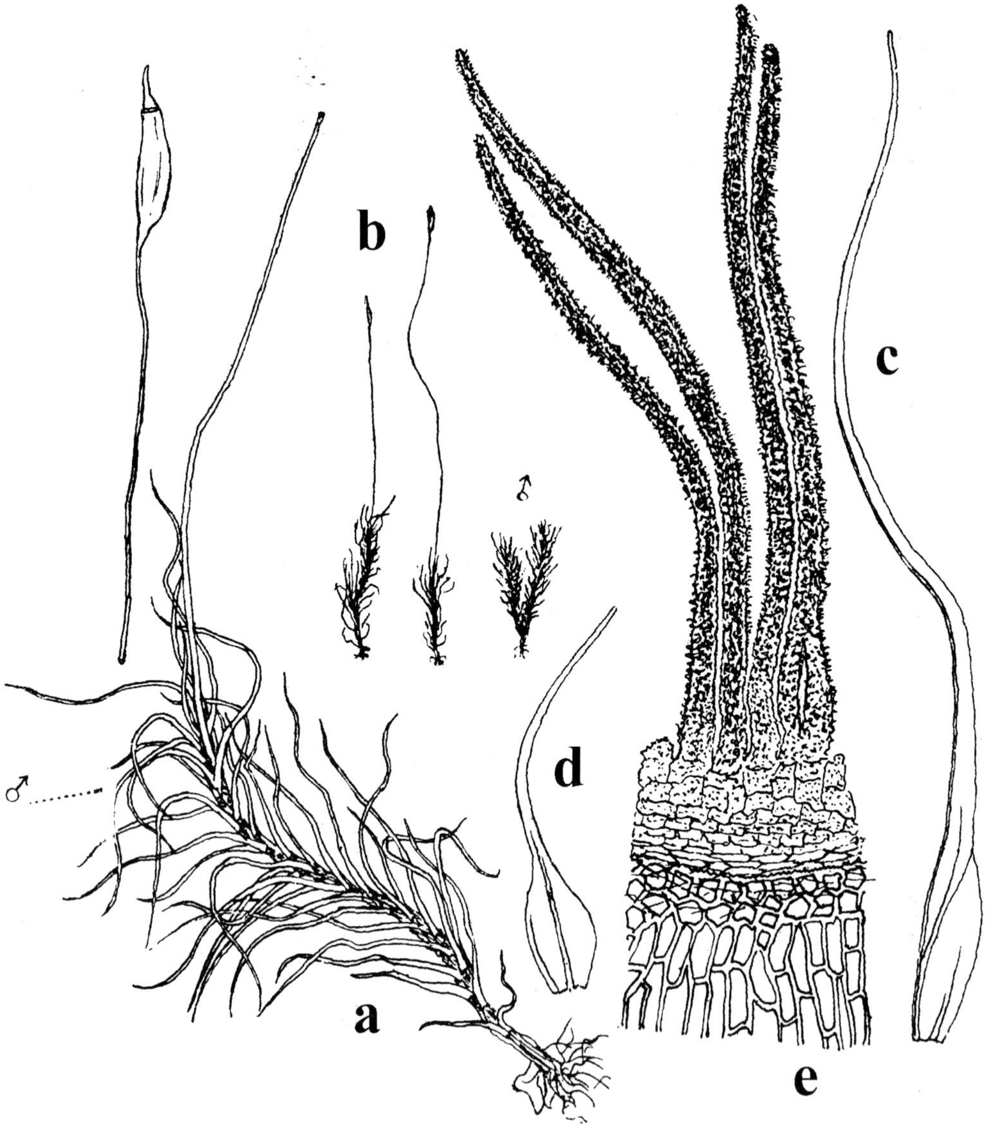

Fig. 15. ***Ditrichum difficile.*** **a.** **Habit. b. Habit female and male plants. c. Upper stem leaf. d. Lower stem leaf. e. Peristome ventral surface. Reproduced from Die Musci der Flora Von Buitenzorg. Max Fleischer, 1902.**

16. Entodon jamesonii (Tayl.) Mitt. (Fig. 16)

Entodon aurescens Hampe
Entodon lutescens Hampe

Plants medium-sized, golden to reddish, glossy, in flat, loose mats. **Stems** prostrate, irregularly branched, the branches complanate-foliate. **Leaves** of stem and branches ± differentiated, erect-spreading, subconcave, lanceolate to broadly-lanceolate, acuminate, 1.25–1.75 mm long, margins plane, serrulate in the upper half; cells linear, not appreciably shorter at the apex. **Setae** yellow, 10–15 mm long; **capsules** ca. 3 mm long; annulus none; exostome teeth reddish, striolate almost to the apex. **Spores** papillose, 19–25 µm diameter.

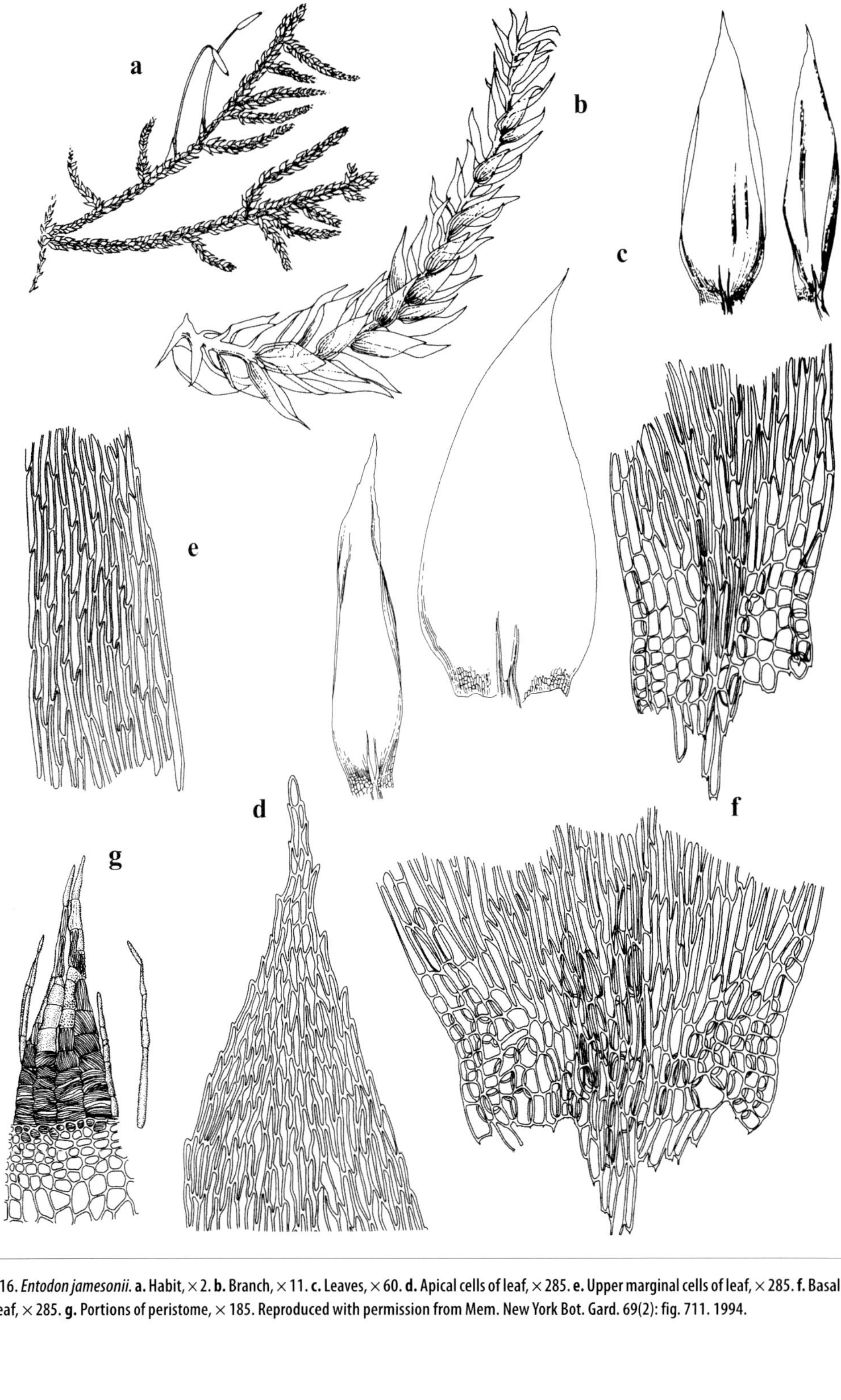

Fig. 16. *Entodon jamesonii.* **a.** Habit, × 2. **b.** Branch, × 11. **c.** Leaves, × 60. **d.** Apical cells of leaf, × 285. **e.** Upper marginal cells of leaf, × 285. **f.** Basal cells of leaf, × 285. **g.** Portions of peristome, × 185. Reproduced with permission from Mem. New York Bot. Gard. 69(2): fig. 711. 1994.

Notes.—This species is distinguished by quadrate alar cells that reach the costa in narrow leaves but not in broader leaves on the same branch. This species was reported by Vargas (1974) but we have not seen the voucher and would not be surprised if it represented the more lowland occurring *Entodon beyrichii* (Schwägr.) Müll. Hal.

Habitat.—Open to forest and shrubby páramo, 1000–3600 m, on soil, rocks, leaf litter, logs, trunk, and branches of shrubs, treelets, and trees.

FISSIDENTACEAE

FISSIDENS Hedw.

Plants small to rather large, yellowish to green or darker, sometimes glossy, gregarious or in loose to compact sods. **Stems** erect or suberect, simple or sparsely branched, radiculose at base. **Leaves** distichous, remote to overlapping; the lower portion of the adaxial side forming two vaginant laminae; ovate- to oblong-lanceolate, obtuse, acute or acuminate; border absent or present throughout or confined to vaginant laminae, rarely confined to the uppermost stem and perichaetial leaves only; **costa** single, mostly stout, subpercurrent to short excurrent; laminal cells usually isodiametric, plane to bulging, mammillose or papillose. **Setae** straight or flexuose; **capsules** erect, inclined or pendulous, symmetrical to asymmetrical; opercula usually long-rostrate; peristome single, teeth 16, deeply forked, reflexed when dry and strongly incurved when moist. **Calyptrae** cucullate or mitrate-rostrate. **Spores** medium-sized, rarely large.

Note.—*Fissidens* was the most diverse genus in our study area and is one of the most diverse groups in the Neotropics. The genus is likely very often neglected by collectors.

1. Leaves not bordered (elimbate).
   2. Leaf cells smooth to bulging ________ **Fissidens pellucidus**
   2. Leaf cells mammilose to unipapillose.
      3. Costa broad, 30–50 µm at apex of vaginant laminae; laminae irregularly bistratose; cells ± isodiametric. ________ **Fissidens allionii**
      3. Costa narrower, 22–40 µm at apex of vaginant laminae; laminae unistratose throughout; cells hexagonal to oblong ________ **Fissidens prionodes**
1. Leaves bordered (limbate).
   4. Leaf cells smooth to bulging.
      5. Leaf costa ending several cells below apex; leaf border 3–4 stratose ________ **Fissidens flaccidus**
      5. Leaf costa percurrent to short excurrent; leaf border 1–2 stratose ________ **Fissidens zollingeri**
   4. Leaf cells papillose to mammilose.
      6. Cells pluripapillose.
         7. Leaf border on lower 1/2 or less of vaginant laminae ________ **Fissidens guianensis**
         7. Leaf border extending the complete length of the vaginant laminae or nearly so ________ **Fissidens intramarginatus**
      6. Cells unipapillose to mammilose.
         8. Leaf border restricted to perichaetial and subtending leaves ________ **Fissidens laegenarius**
         8. Leaf border present on all leaves.
            9. Leaf border not extending more than 0.75 the length of the vaginant laminae; vaginant laminae very broad, appearing to encircle the stem ________ **Fissidens perfalcatus**
            9. Leaf border extending to the apices of vaginant laminae, often onto the adjacent ventral laminae ________ **Fissidens submarginatus**

17. Fissidens allionii Broth. (**Fig. 17**)

Plants small, mostly 4–6 mm tall. **Leaves** to 1.5 x 0.2 mm, acute, vaginant laminae unequal; margins entire to serrulate, not bordered; **costa** broad, 30–50 µm wide at apex of vaginant laminae, occupying 1/2 or more the width of the leaves, excurrent; cells irregularly bistratose, ± isodiametric, 6–9 µm in diameter, ± sharply mammilose-unipapillose. **Sporophytes** terminal on short axillary branches.

Notes.—*Fissidens allionii* is recognized by its large percurrent to short-excurrent costa that occupies 1/2 or more the width of the narrow leaf. Moreover, the leaves are tightly imbricate and the mammilose laminal cells are bistratose in places.

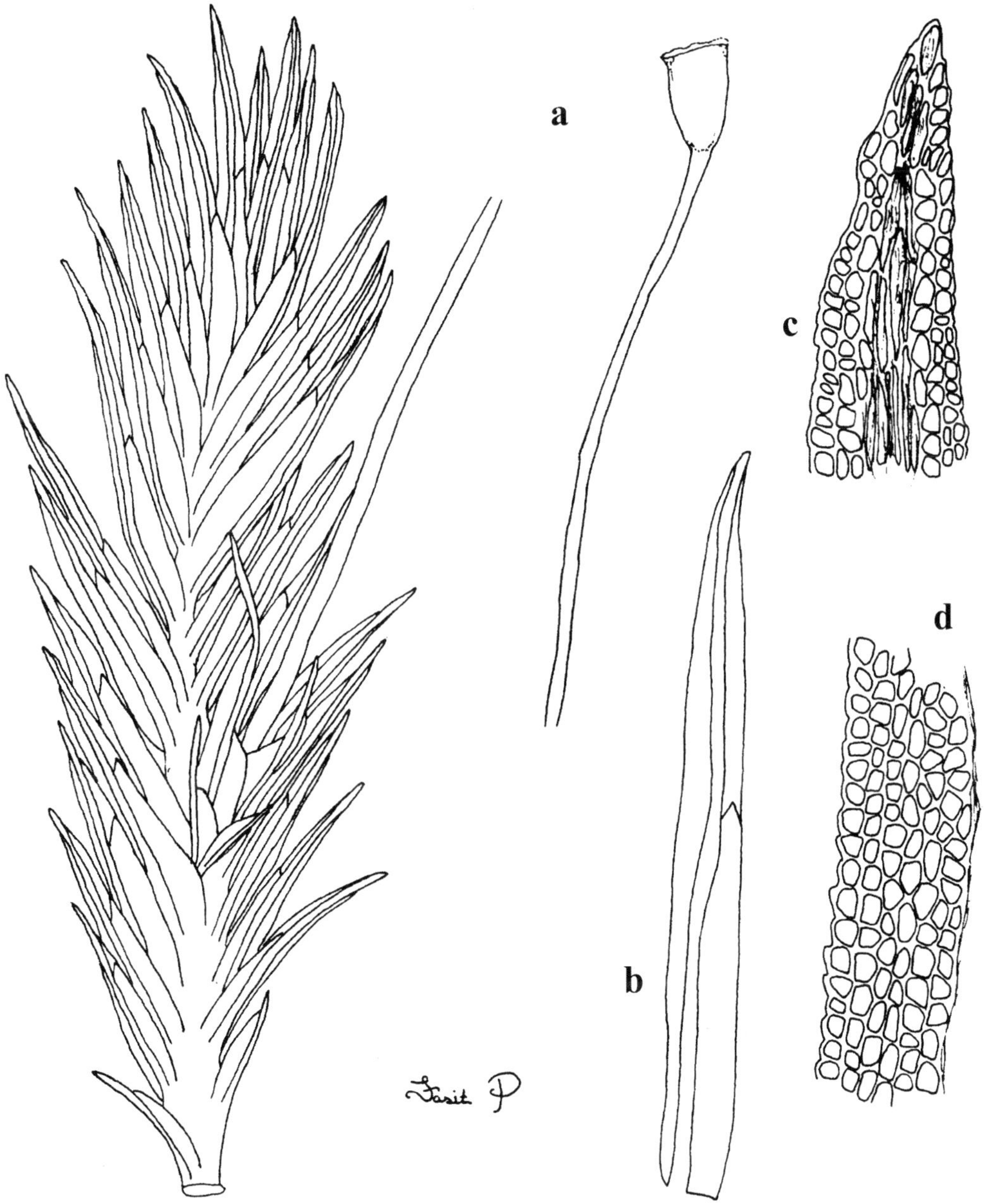

Fig. 17. *Fissidens allionii.* **a.** Habit. **b.** Leaf. **c.** Leaf apex. **d.** Mid dorsal lamina. Drawn by Ronald Pursell. Reproduced with permission from Mem. New York Bot. Gard. 76(3): fig. 3. 2003.

Habitat.—Wet lowland to lower montane forest, 120–1200 m, on soil.

18. Fissidens flaccidus Mitt. (**Fig. 18**)

Fissidens mollis Mitt.

Plants medium-sized, to ca. 15 mm tall. **Leaves** flaccid, 4–5.5 × 0.5–0.75 mm, acute to obtuse-apiculate, vaginant laminae ± equal; margins often serrulate above, entire below, bordered on all laminae with limbidia

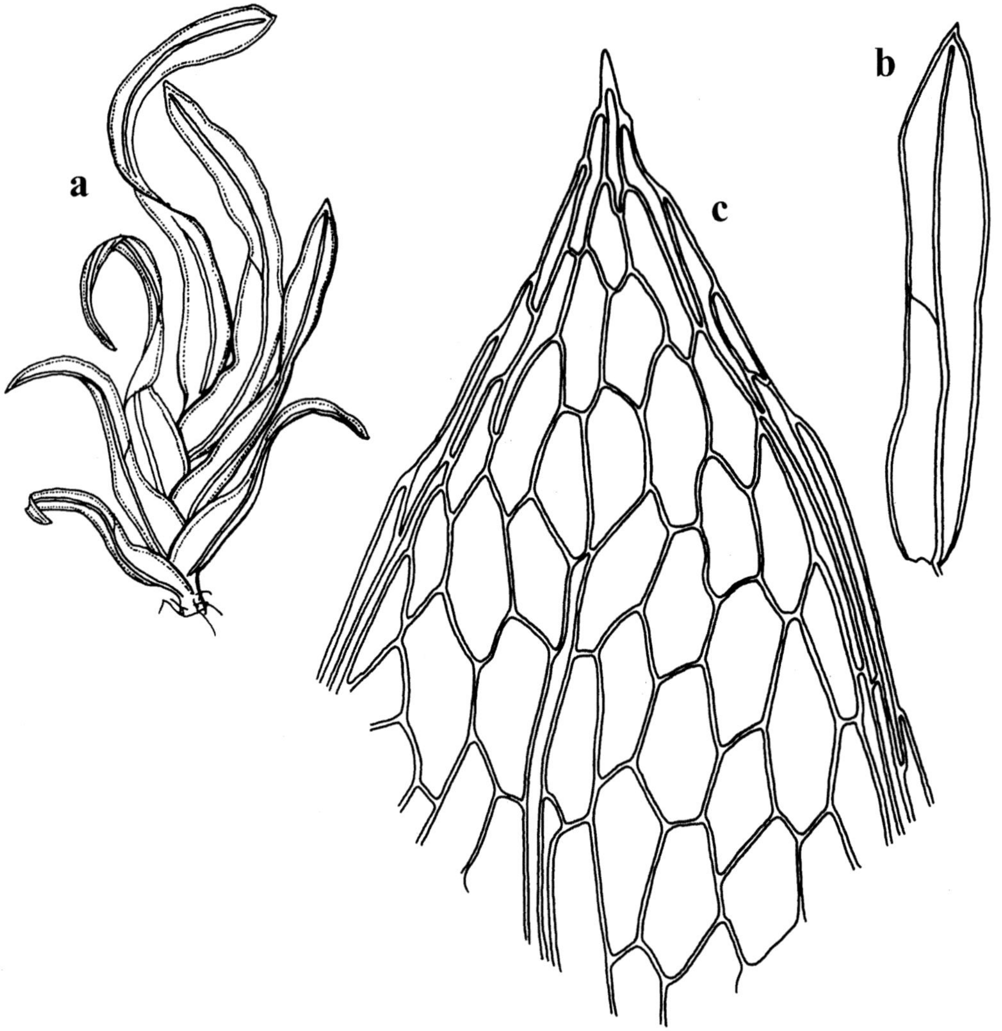

Fig. 18. *Fissidens flaccidus.* **a.** Habit. **b.** Leaf. **c.** Leaf apex. Reproduced with permission from Mem. New York Bot. Gard. 76(3): fig. 6. 2003.

confluent at leaf apex, 3–4 stratose; **costa** subpercurrent to percurrent; cells quadrate to hexagonal, 20–40 × 14–19 µm, smooth. **Sporophytes** terminal.

Habitat.—Lowland to montane forest, 10–2800 m, on flooded soil or rocks submerged in streams.

19. Fissidens guianensis Mont. (Fig. 19)

Plants small, 2.5–5 mm tall. **Leaves** ca. 1.3 × 0.3 mm, obtuse or rounded-acute, vaginant laminae unequal; margins crenulate-serrulate, bordered in lower 1/2 or less of most vaginant laminae; **costa** subpercurrent; cells rounded-hexagonal, 4–6 µm in diameter, pluripapillose. **Sporophytes** terminal.

Notes.—The only other pluripapillose species in the study area is *F. intromarginatus*, which differs from *F. guianensis* in having the border extending the complete length of the vaginant laminae.

Habitat.—Lowland non-flooded moist forest, 20–1600 m, on soil, termite mounds, and the base of small trees. *Fissidens guianensis* is a very common species of the lowland rainforest.

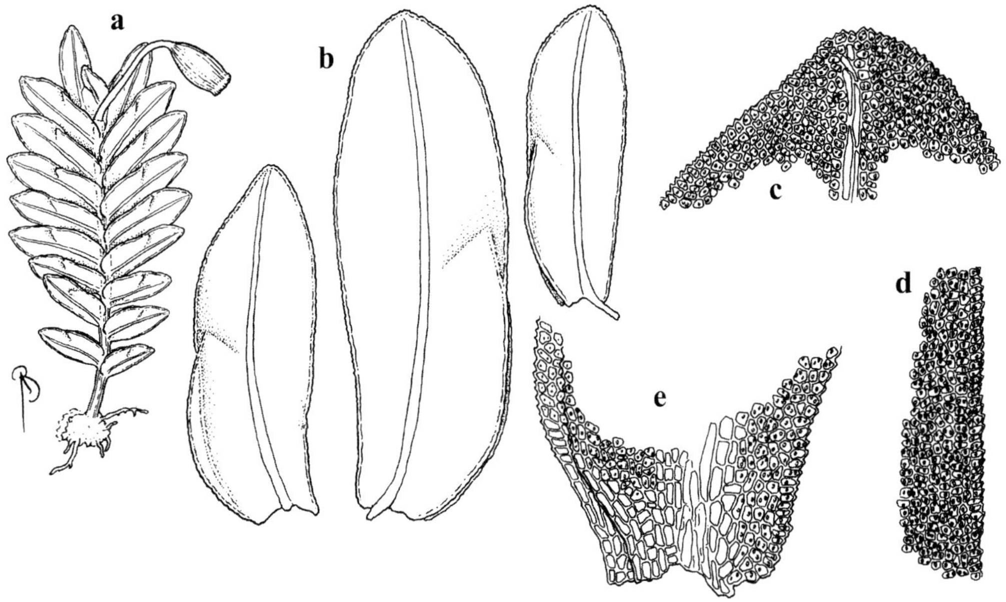

Fig. 19. *Fissidens guianensis*. a. Habit. b. Leaves. c. Leaf apex. d. Median marginal cells of ventral lamina. e. Leaf base. Drawn by Bobbi Angell. Reproduced with permission from Mem. New York Bot. Gard. 76(3): fig. 7. 2003.

20. Fissidens intramarginatus (Hampe) A. Jaeger (Fig. 20)

Fissidens intramarginatus (Hampe) Mitt.
Fissidens tenuifolius Mitt.

Plants medium-sized, 5–14 mm tall. **Leaves** 0.9–1.5 × 0.3 mm, apex acute, vaginant laminae equal, ca. 1/3 the leaf length; margins crenulate-serrulate, bordered on all vaginant laminae, extending onto lower part of ventral laminae; **costa** percurrent to short excurrent; cells rounded-hexagonal, 4–7 µm, pluripapillose. **Sporophytes** terminal.

Habitat.—Lowland to montane forest, 10–3050 m, on soil and limestone rocks and boulders, often on stream banks and bluffs.

21. Fissidens lagenarius Mitt. var. muriculatus (Mitt.) Pursell (Fig. 21)

Fissidens diplodus Mitt.

Plants small, to ca. 8 mm tall. **Leaves** 1 × 0.4 mm, acute to apiculate, vaginant laminae ± equal; margins crenulate-serrulate, limbate 1/3 to 1/2 of length of vaginant laminae of perichaetial leaves and those immediately subtending; **costa** subpercurrent to percurrent; cells rounded to hexagonal, 6–10 µm in diameter, with long, simple or forked papillae. **Sporophytes** terminal.

Habitat.—Lowland to montane forest, 70–2400 m, on the base of trees and logs.

22. Fissidens pellucidus Hornsch. (Fig. 22)

Plants small, to ca. 8 mm tall. **Leaves** ca. 1.7 × 0.3 mm, acute to bluntly acute, vaginant laminae unequal; margins entire to crenulate-serrulate, border restricted to vaginant laminae of perichaetial leaves and those immediately subtending; **costa** percurrent to short-excurrent; cells hexagonal, 11–20 µm in diameter, smooth to bulging. **Sporophytes** terminal.

Notes.—*Fissidens pellucidus* is the only bordered, non-papillose species in the study area. The plants are

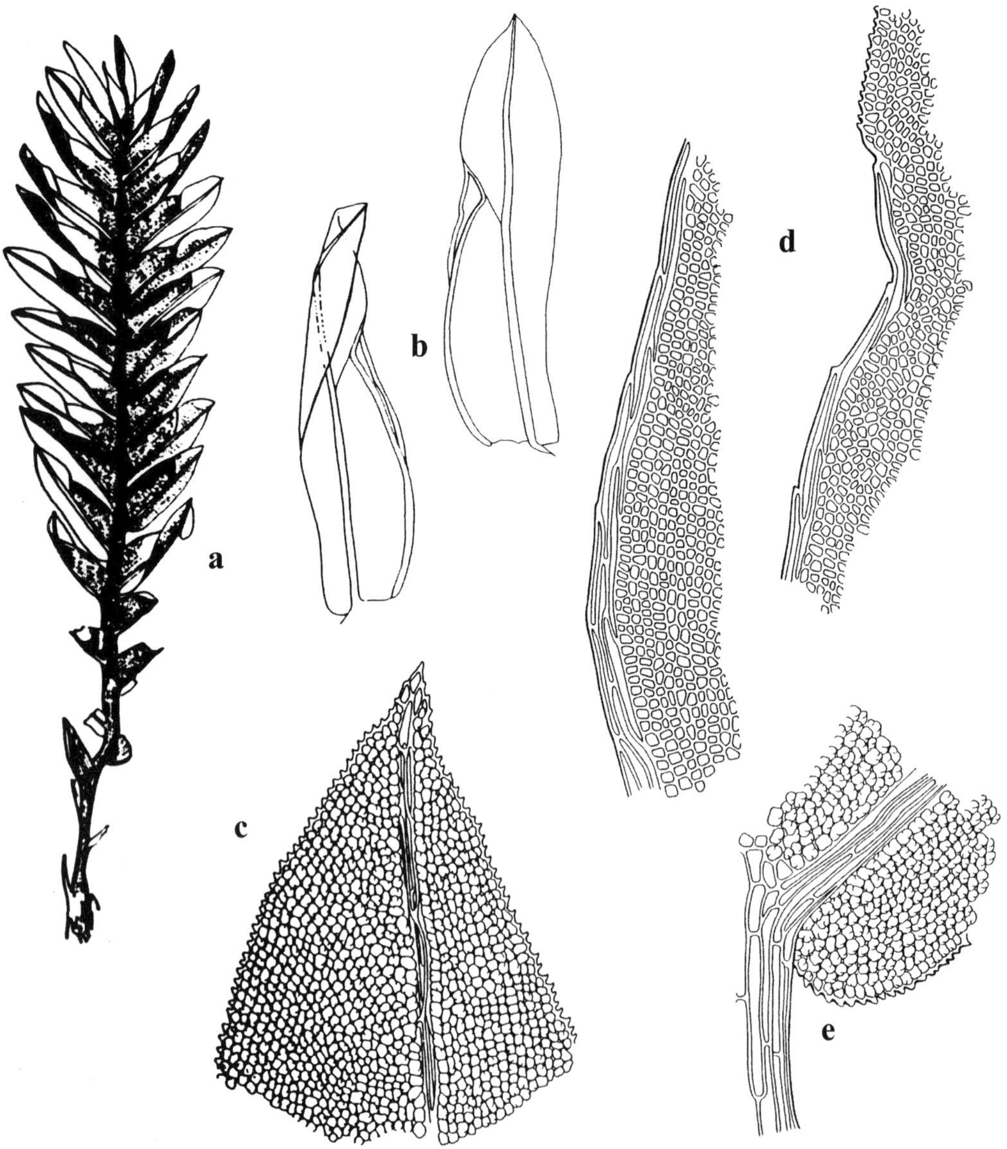

FIG. 20. *Fissidens intramarginatus*. **a.** Habit, × 26. **b.** Leaves, × 78. **c.** Apical cells of leaf, × 370. **d.** Marginal cells of vaginant lamina (without indication of papillae), × 370. **e.** Base of dorsal lamina, × 370. Reproduced with permission from Mem. New York Bot. Gard. 69(1): fig. 32. 1994.

generally light green and red-tinged in the costa and stem. The smooth laminal cells contain guttulae (cell lumen rounded because of strong wall thickening) that can be mistaken for papillae.

Habitat.—Lowland to lower montane forest, 10–2200 m, mostly on bases of trees and small shrubs, less often on rock, soil, and termite mounds.

23. Fissidens perfalcatus Broth. (**Fig. 23**)

Plants small, to 5 mm tall. **Leaves** 0.8 × 0.3 mm; apex acute; vaginant laminae slightly unequal; margins entire and bordered on vaginant laminae for 1/3–3/4 the leaf laminae, crenulate-serrulate on ventral and

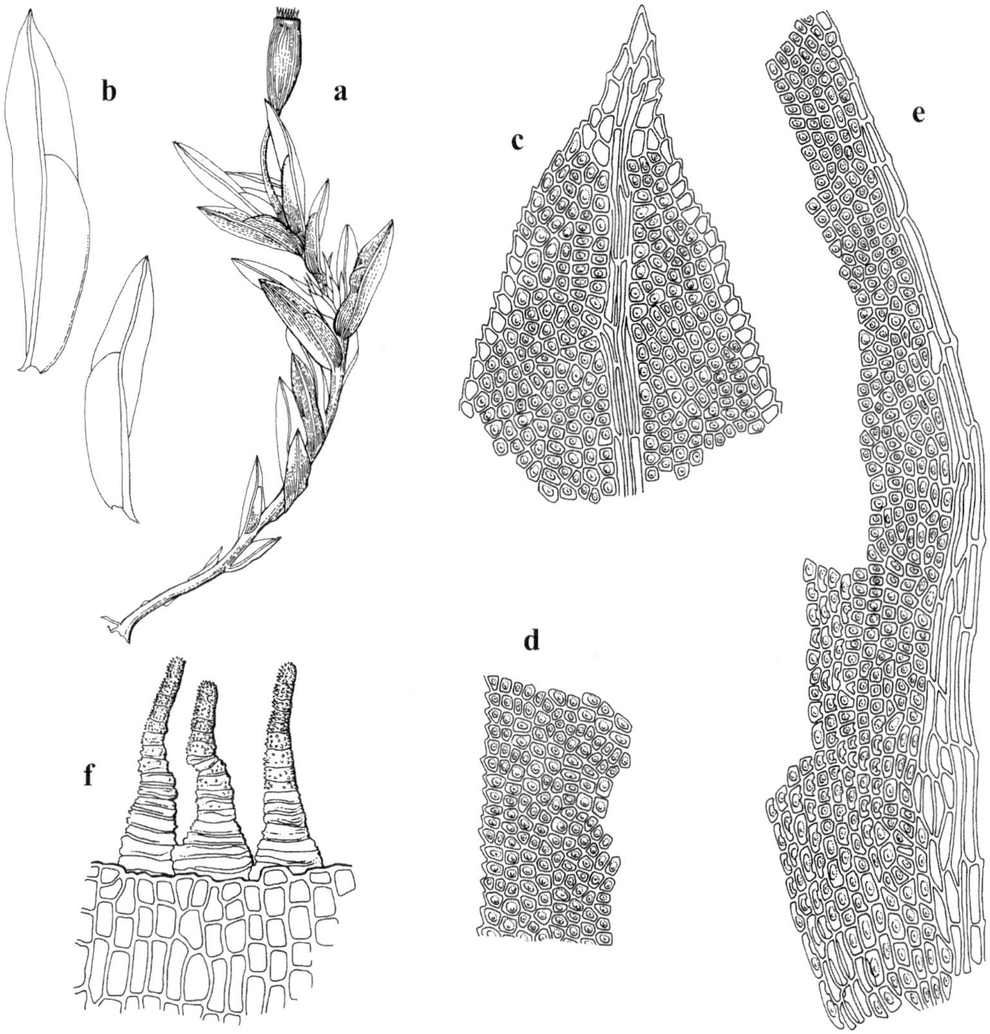

FIG. 21. *Fissidens lagenarius* var. *muriculatus*. **a.** Habit. **b.** Leaves. **c.** Leaf apex. **d.** Marginal cells of apical lamina. **e.** Base of vaginant lamina. **f.** Portion of Peristome. Reproduced with permission from Mem. New York Bot. Gard. 76(3): fig. 10. 2003.

dorsal laminae; **costa** percurrent to short excurrent; cells quadrate to irregularly hexagonal, 7–12 µm in diameter, sharply unipapillose. **Sporophytes** terminal.

Notes.—Fissidens perfalcatus has broadly triangular, acute stem leaves with wide spreading, recurved vaginant laminae that nearly encircle the stem. The border is restricted to the lower 3/4 of the vaginant laminae, except on perichaetial leaves, where it extends the entire length of the vaginant laminae.

Habitat.—Lowland to lower montane forest, 200–930 m, partly shaded banks on clay soil.

24. Fissidens prionodes Mont. (Fig. 24)

Plants small, to 5 mm tall, often ± reddish. **Leaves** 1–2 × 0.1–0.35 mm, acute to acuminate, vaginant laminae ± unequal; margins crenulate-serrulate to serrate, not bordered; **costa** percurrent to short-excurrent, ca. 22–40 µm wide at apex of vaginant laminae; cells hexagonal to oblong, 10–25 µm, mammilose-unipapillose, unistratose throughout. **Sporophytes** terminal on short axillary branches.

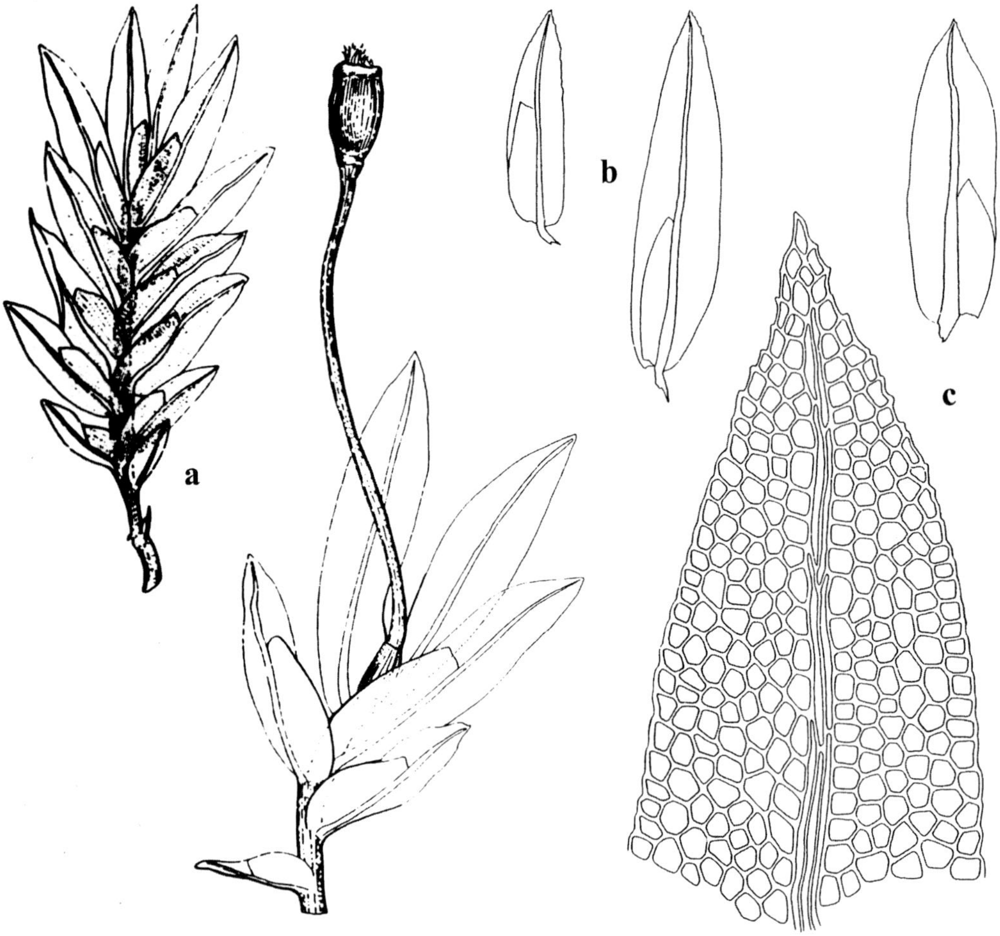

Fig. 22. ***Fissidens pellucidus.*** **a. Habits, × 41 & 81. b. Leaves, × 81. c. Apical cells of leaf, × 309. Reproduced with permission from Mem. New York Bot. Gard. 69(1): fig. 28. 1994.**

Notes.—The only other borderless, unipapillose species in the study area is *F. allionii*, which generally has a wider costa than *F. prionodes*, as well as irregularly bistratose laminae.

Habitat.—Moist lowland forest, rare at higher elevations, 80–2900 m, on soil and termite mounds.

25. **Fissidens submarginatus** Bruch (**Fig. 25**)

Plants small, to ca. 2.5 mm. **Leaves** 1 × 0.35 mm; apex acute, vaginant laminae ± equal; margins serrulate-crenulate, entire to slightly denticulate on the vaginant laminae, bordered along the entire length of the vaginant laminae; **costa** percurrent to ending 2–3 cells below the apex; cells quadrate to hexagonal, 5–7 µm, unipapillose. **Sporophytes** terminal.

Notes.—*Fissidens submarginatus* is related to *F. perfalcatus*. In *F. perfalcatus*, the vaginant laminae are very broad and nearly encircle the stem. The border on most leaves of *F. submarginatus* usually extends the length of the vaginant laminae and may run onto the adjacent ventral laminae. In *F. perfalcatus*, the border on all leaves, except the perichaetial ones, typically ends below the distal ends of the vaginant laminae.

Habitat.—Non-flooded lowland to montane forest, 35–2400 m, on rocks or soil.

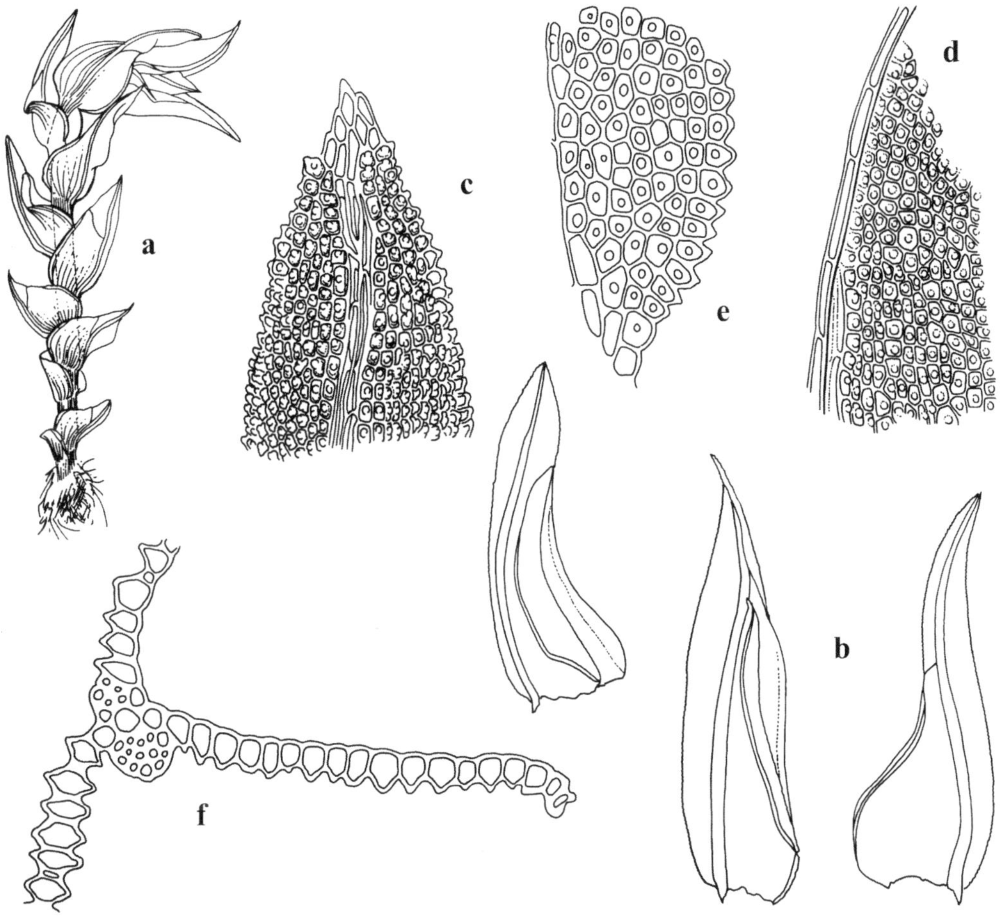

FIG. 23. *Fissidens perfalcatus.* **a.** Habit, × 44. **b.** Leaves, × 85. **c.** Apical cells of leaf, × 337. **d.** Margin of vaginant lamina, × 337. **e.** Base of dorsal lamina, × 804. **f.** Cross-section near base of leaf, × 804. Reproduced with permission from Mem. New York Bot. Gard. 69(1): fig. 48. 1994.

26. Fissidens zollingeri Mont. (**Fig. 26**)

Plants small, to 5.5 mm tall; sometimes with clavate propagula in the leaf axils. **Leaves** to 2.5 × 0.5 mm, acute to apiculate to acuminate, vaginant laminae equal, sometimes with clavate propagula in the leaf axils; margins ± entire, bordered on all laminae, 1–2 stratose; **costa** percurrent to short-excurrent; cells hexagonal, 10–15 µm, smooth to somewhat bulging, those juxtacostally in proximal part of vaginant laminae enlarged, oblong and pellucid. **Sporophytes** terminal.

Notes.—Typically the plants are pale green and frond-like. The only other non-papillose, bordered species in the study area is *F. flaccidus* which is easily separated from *F. zollingeri*, by, among other things, the cell size difference and the multistratose (3–4 stratose) limbidium of *F. flaccidus* versus the 1–2 stratose limbidium of *F. zollingeri*.

Habitat.—Non-flooded lowland to montane forest, 20–2750 m, on soil and rocks; frequent along stream banks. A widely distributed species in neotropical lowlands.

HYPNACEAE

Plants small to robust, in loose or dense mats. **Stems** mostly creeping, sometimes ascending, irregularly

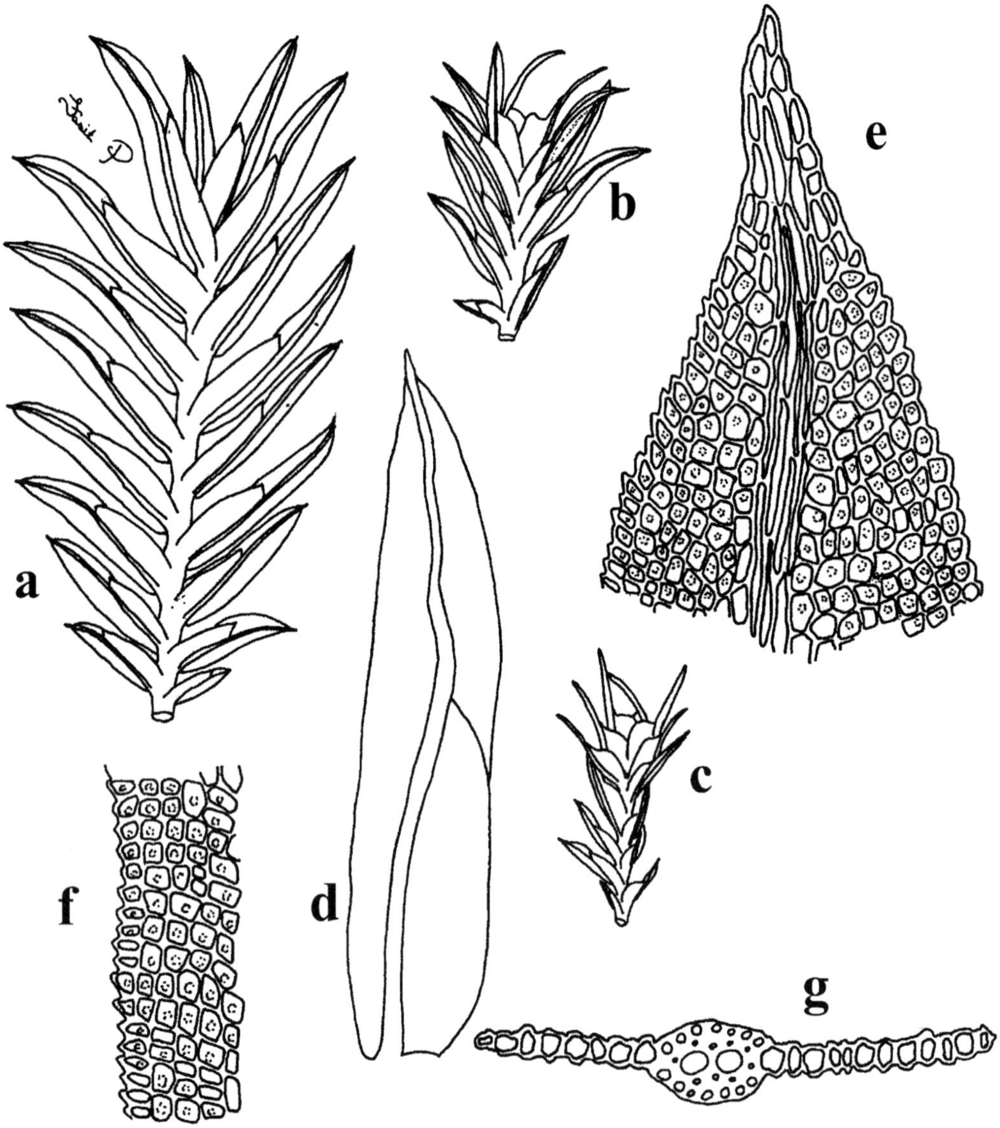

FIG. 24. *Fissidens prionodes*. **a.** Habit of sterile stem. **b, c.** Habits of male plants. **d.** Leaf. **e.** Leaf apex. **f.** Marginal dorsal laminal cells. **g.** Cross-section of apical lamina. Drawn by Ronald Pursell. Reproduced with permission from Mem. New York Bot. Gard. 76(3): fig. 17. 2003.

branched to very regularly pinnate, often homomallous to secund or falcate-secund, occasionally somewhat complanate. **Leaves** of stems and branches mostly uniform, sometimes ± differentiated and ± asymmetrical, generally ovate to ovate-lanceolate, often-acuminate; **costa** short and double or lacking; cells mostly linear, smooth or occasionally papillose because of projecting cell angles (rarely with several papillae in a row over the lumina at back), mostly differentiated at basal angles, usually small and quadrate but sometimes moderately enlarged and inflated in rather small but usually distinct alar groups. **Setae** elongate, mostly smooth; **capsules** erect or, more commonly, inclined to horizontal, ovoid to cylindrical, straight or curved and asymmetrical; annulus usually differentiated; opercula ± conical or sometimes convex, often rostrate;

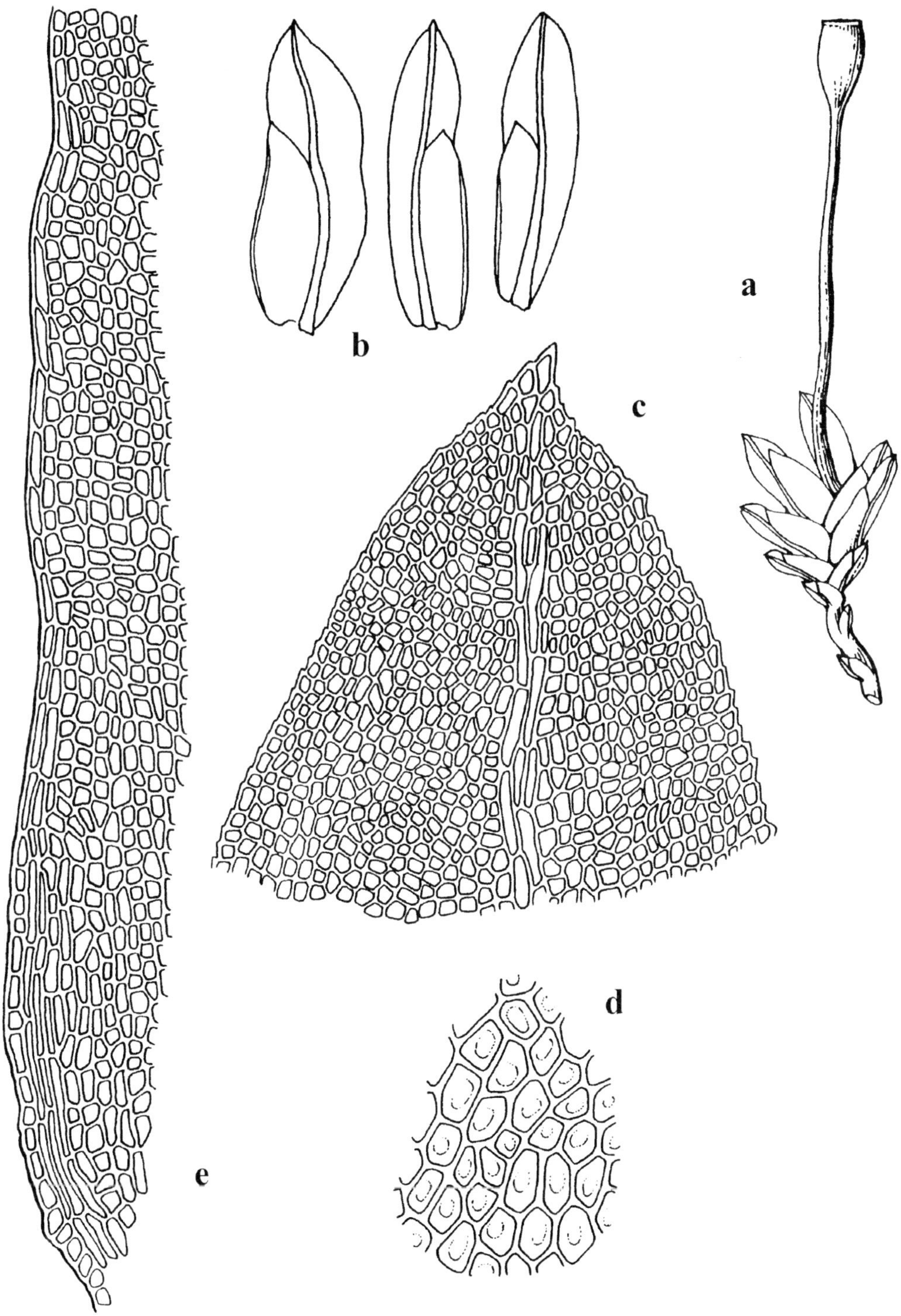

FIG. 25. *Fissidens submarginatus.* **a.** Habit, × 22. **b.** Leaves, × 54. **c.** Apical cells of leaf (without indication of papillae), × 412. **d.** Apical cells (with indication of papillae) × 982. **e.** Cells of vaginant lamina, × 412. Reproduced with permission from Mem. New York Bot. Gard. 69(1): fig. 49. 1994.

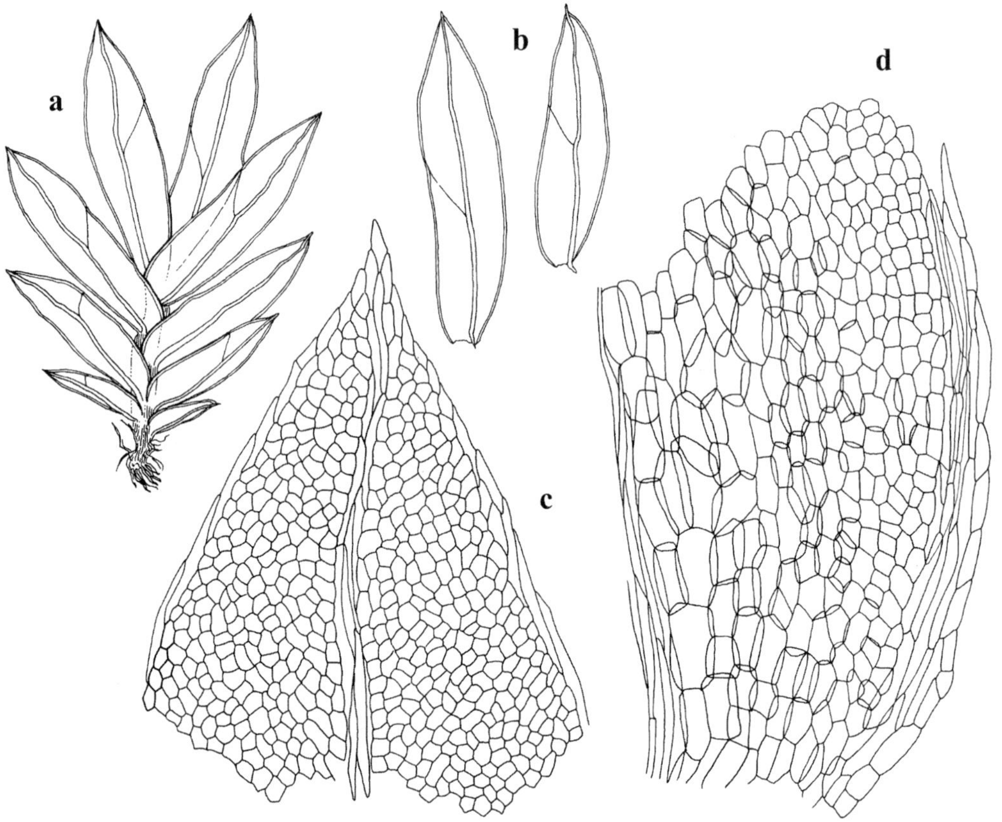

Fig. 26. *Fissidens zollingeri.* **a.** Habit, × 27. **b.** Leaves, × 30. **c.** Apical cells of leaf, × 309. **d.** Base of vaginant lamina, × 309. Reproduced with permission from Mem. New York Bot. Gard. 69(1): fig. 51. 1994.

peristome double (or rarely the endostome lacking), typically well-developed and consisting of 16 lanceolate-subulate teeth with fine cross striolations on the outer surface. **Calyptrae** cucullate, smooth, naked or rarely ± hairy. **Spores** spherical.

1. Leaf cells smooth or inconspicuously and irregularly prorulose (cells projecting at angles).
   2. Branch leaves markedly dimorphic (lateral and ventral leaves differentiated) ____________ **Rhacopilopsis trinitensis**
   2. Branch leaves not or scarcely differentiated.
      3. Leaf areolation very lax; median to upper cells short to oblong hexagonal ______________ **Vesicularia vesicularis**
      3. Leaf cells narrowly prosenchymatous; median to upper cells linear ___________________ **Isopterygium tenerum**
1. Leaf cells conspicuously prorulose.
   4. Plants stipitate, with arched stem and slender, tapering branches; laminal cells projecting at upper angles ____ **Mittenothamnium reptans**
   4. Plants not stipitate, prostrate, with ± blunt branches; laminal cells projecting at upper and lower angles ____ **Chryso-hypnum diminutivum**

CHRYSO-HYPNUM (Hampe) Hampe

Plants mostly small, sometimes medium-sized, in soft, often extensive, dense mats. **Stems** creeping, prostrate, freely and irregularly to regularly pinnately branched, the branches mostly simple, short, blunt, and densely foliate. **Stem and branch leaves** similar, erect to wide-spreading, mostly ovate, abruptly short to long-acuminate; margins serrulate; **costa** short and double; cells linear, prorulose at upper and lower angles at back, rarely smooth, becoming short and broader toward the insertion; alar cells few in extreme angles,

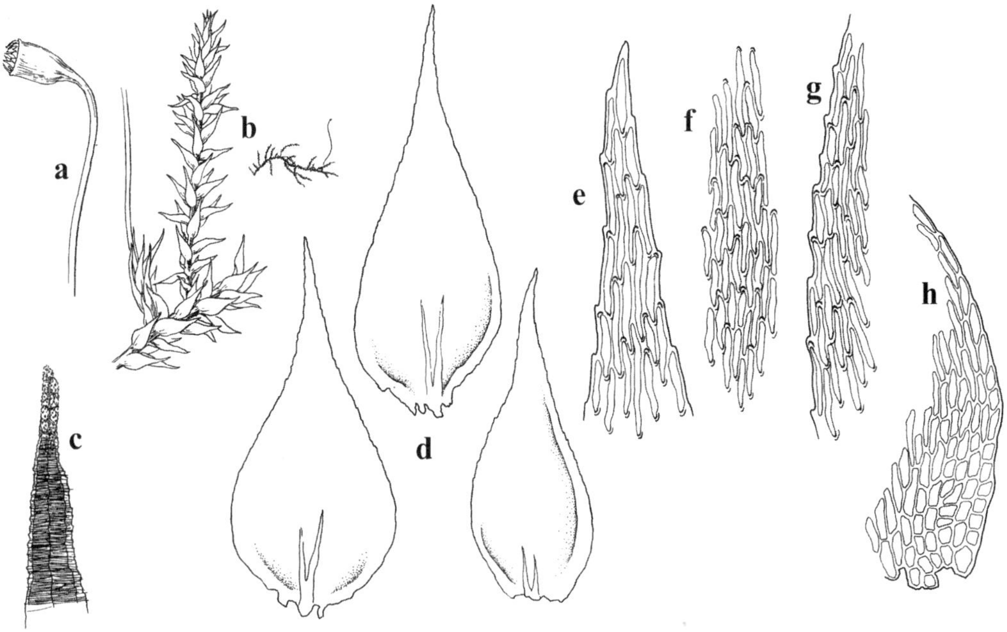

FIG. 27. *Chryso-hypnum diminutivum*. **a.** Portion of stem and sporophyte. **b.** Habit. **c.** Exostome tooth. **d.** Leaves. **e.** Leaf apex. **f.** Laminal cells, midleaf. **g.** Leaf margin. **h.** Basal marginal leaf cells. Reproduced with permission from Mem. New York Bot. Gard. 82: plate 132, figs. 1–8. 1998.

quadrate to short-rectangular. **Setae** elongate, smooth, reddish; **capsules** horizontal to pendent, ± curved, somewhat asymmetrical, mostly ovoid to short-cylindrical, often constricted below the mouth when dry; annulus differentiated; opercula conical, mostly apiculate; peristome double. **Calyptrae** cucullate, mostly sparsely to densely hairy, rarely naked, smooth. **Spores** finely papillose.

27. Chryso-hypnum diminutivum (Hampe) W.R. Buck (**Fig. 27**)

Plants mostly pale- to dark-green, sometimes golden or brown-tinged. **Stem leaves** erect to wide-spreading, ovate, 0.8–1.2 m long, gradually and narrowly acuminate; margins serrulate ± throughout, plane; **costa** short and double, unequal, one fork often ending 1/3–1/2 the leaf length; cells linear, prorulose at upper and lower ends at back. **Branch leaves** usually wide-spreading, lanceolate to ovate-lanceolate, 0.6–0.9 mm long, gradually acuminate, concave; margins serrulate almost to base; **costa** short and double, unequal, one fork often ending 1/4 to 1/3 the leaf length; cells linear to oblong-linear, conspicuously prorulose at upper and lower ends at back. **Setae** 8–25 mm long; **capsules** 0.7–1.1 mm long; exostome teeth shouldered, bordered, on the front surface cross-striolate below, coarsely papillose above; endostome segments finely papillose, perforate, ca. as long as the teeth. **Calyptrae** naked or with a few hairs. **Spores** 11–19 µm.

Notes.—The often dense mats of smallish, prostrate, subcomplanate-foliate plants with ovate leaves whose cells are prorulose at both upper and lower ends will readily distinguish *Chryso-hypnum diminutivum*.

Habitat.—Moist to semi-dry lowland forest to montane forest, 180–2600 m, on soil, leaf litter, logs, and rocks. Widespread in the tropics.

ISOPTERYGIUM Mitt.

Plants small to medium-sized, in thin to dense mats. **Stems** creeping, almost simple or irregularly but freely branched, usually complanate. **Stem and branch leaves** similar, erect to erect-spreading to wide-spreading, symmetrical or asymmetrical, lanceolate to ovate, acute to acuminate; margins entire to serrulate

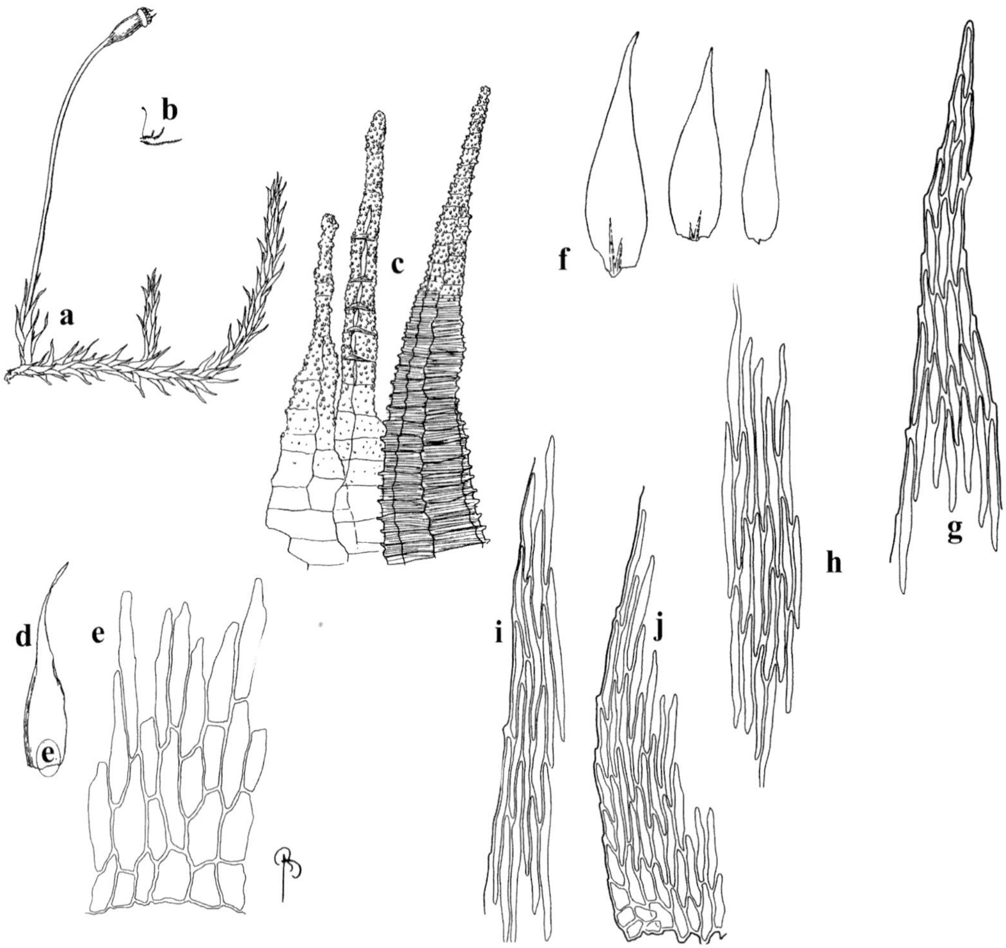

FIG. 28. *Isopterygium subbrevisetum*. **a.** Portion of stem and sporophyte. **b.** Habit. **c.** Leaves. **d.** Leaf apex. **e.** Leaf margin. **f.** Perichaetial leaf. **g.** Perichaetial leaf basal cells. **h.** Portion of peristome. Reproduced with permission from Mem. New York Bot. Gard. 82: plate 125, figs. 9–16. 1998.

above, subentire to entire below; **costa** short and double or absent; cells linear, often flexuose, smooth, thin- to firm-walled, not porose, not becoming shorter in the acumen, becoming shorter, ± rectangular and sometimes porose toward the insertion; alar cells differentiated, few, quadrate to short-rectangular. **Setae** elongate, smooth, reddish; **capsules** erect to inclined or pendent, straight to curved, cylindrical to short-cylindrical, constricted below the mouth when dry; annulus not differentiated; opercula conical to short conical-rostrate; peristome double, coarsely papillose, trabeculate at back. **Calyptrae** cucullate, naked, smooth. **Spores** spherical, smooth to papillose.

28. Isopterygium tenerum (Sw.) Mitt. (Fig. 28)

Plants small to medium-sized, pale- to yellow-green. **Stems** creeping, irregularly but freely branched, ± complanate-foliate. **Stem and branch leaves** similar, erect-spreading, lanceolate to ovate-lanceolate, 0.5–1.4 mm long, gradually acuminate, often long-acuminate, the apex sometimes twisted; margins entire, sometimes minutely serrulate, plane; **costa** short and double, often absent; cells linear, ± straight, smooth, thin-walled, not porose, becoming shorter, broader and laxer in 1–3 rows across the insertion; alar cells few but often conspicuous, in basal angles, quadrate to subquadrate. **Setae** elongate, smooth, orange to reddish, 5–2.2 mm long; **capsules** inclined to pendent, ± arcuate, asymmetrical, short-cylindrical, constricted below the mouth when dry; exostome teeth shouldered, bordered, on the front surface cross-striolate below, coarsely papillose above. **Spores** finely papillose, 9–24 µm in diameter.

Habitat.—Lowland to montane forest, 35–2600 m, on most substrates, especially decaying wood and the bases of trees in a wide range of habitats.

MITTENOTHAMNIUM Henn.

Plants small to medium-sized, in dark green to golden, often extensive, loose mats. **Stems** stipitate, mostly arched, usually wiry, radiculose at tips, freely and irregularly branched, sometimes subcomplanate-foliate. **Stem leaves** often differentiated, mostly wide-spreading to squarrose, broadly ovate, gradually or ± abruptly acuminate, stem and branch leaves similar but stem leaves larger; margins serrate to serrulate; **costa** short and double, sometimes to 1/3 the leaf length, often with one fork longer than the other; cells linear, prorulose at upper ends at back of usually some cells but rarely all, rarely smooth; alar cells few in extreme angles, quadrate to subquadrate. **Setae** elongate, reddish; capsules horizontal to pendent, curved, asymmetrical, cylindrical, usually with a differentiated neck, often constricted below the mouth when dry; annulus differentiated; opercula short to medium conical-rostrate; peristome double, exostome teeth shouldered, bordered, on the front surface cross-striolate below, papillose above. **Calyptrae** cucullate, naked or with a few scattered hairs, smooth. **Spores** spherical, finely papillose.

29. Mittenothamnium reptans (Hedw.) Cardot (**Fig. 29**)

Plants small to medium-sized, in green to golden, extensive, loose mats. **Stems** stipitate, arched, wiry, radiculose at tips, freely and irregularly branched above sparsely foliate stipe. **Stipe leaves** differentiated, widely spaced, wide-spreading, ovate, 0.6–0.8 mm long, gradually short-acuminate, somewhat decurrent; margins serrulate almost to base; cells long-hexagonal to linear-oblong, smooth or sparsely prorulose at upper ends at back, becoming rectangular toward the insertion; alar cells few, short-rectangular in extreme angles. **Stem leaves** erect- to wide-spreading, lanceolate to ovate-lanceolate, 0.45–1.3 mm long, gradually short-acuminate; margins serrate to serrulate above, serrulate to denticulate below; cells linear, sparsely prorulose at upper ends at back, only becoming shorter at extreme insertion; alar cells very few in 1–4 rows in extreme angles. **Branch leaves** laxly spaced, erect- to wide-spreading, lanceolate, 0.4–1.0 mm long, gradually short-acuminate, margins mostly serrate above, serrulate almost to base; cells long-hexagonal to linear-oblong, sparsely but usually conspicuously prorulose at upper ends at back, only becoming shorter at extreme insertion; alar cells quadrate to short-rectangular in 3–5 rows in extreme angles. **Setae** elongate, dark-red, 1–2 cm long; **capsules** inclined to pendent, arcuate, asymmetrical, cylindrical, 1.2–1.5 mm long, constricted below the mouth when dry; annulus of 1–2 rows of oval to rectangular thick-walled cells; exostome teeth shouldered, bordered, on the front surface cross-striolate below, coarsely papillose above, trabeculate at back. **Spores** 10–17 µm.

Notes.—*Mittenothamnium reptans* is recognizable by its arched, sprawling stems. The plants are laxly foliate and the young leaves appear particularly serrate due to the serrations that develop before full leaf size.

Habitat.—Mostly from lower to upper montane forest and open areas at forest margins in secondary sites; infrequent in lowland forests, 150–3400 m, found growing on a wide range of substrates, including soil, rocks, and trees.

RHACOPILOPSIS Renauld & Cardot

Plants mostly small and slender but elongate, or medium-sized, in often lustrous, soft, green to pale-green, thin mats. **Stems** creeping, irregularly to regularly pinnate, the branches complanate-foliate. **Stem and branch leaves** somewhat differentiated; **ventral, dorsal, and lateral stem leaves** sometimes differentiated; **stem leaves** with margins subentire to distantly subserrulate throughout, plane to recurved; **costa** short

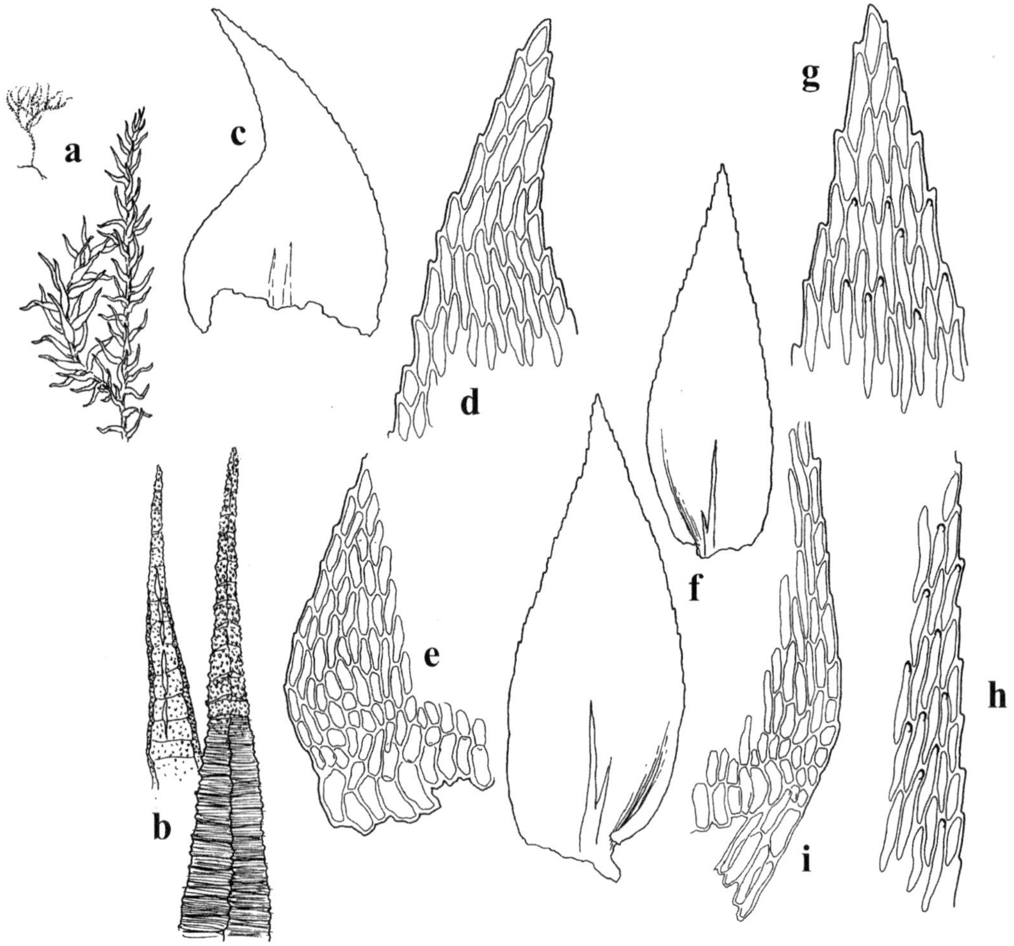

Fig. 29. *Mittenothamnium reptans.* **a.** Habit and portion of stem. **b.** Portion of peristome. **c.** Stem leaf. **d.** Stem leaf apex. **e.** Basal marginal stem leaf cells. **f.** Branch leaves. **g.** Branch leaf apex. **h.** Branch leaf margin. **i.** Basal marginal leaf cells. Reproduced with permission from Mem. New York Bot. Gard. 82: plate 131, figs. 11–19. 1998.

and double; cells long-hexagonal to linear, smooth but with thickened cell end walls; alar cells differentiated in extreme basal angles, 2–4 cells enlarged and inflated, ± oblong, sometimes colored, with a small group of subquadrate to short-rectangular cells directly above them; in lateral leaves the enlarged cells sometimes only on one side of the costa.

Notes.—Branch leaves with dorsal and lateral leaves usually differentiated help in recognition of this genus.

30. Rhacopilopsis trinitensis (C. Müll.) E.Britton & Dixon in Dixon (**Fig. 30**)

Plants slender to medium-sized in mostly pale-green, thin mats. **Stems** creeping, to 10 cm long, the branches complanate-foliate, less than 1 cm long. **Stem and branch leaves** somewhat differentiated; **ventral, dorsal, and lateral stem leaves** strongly differentiated, **ventral leaves** symmetrical, oblong-ovate, 0.65–0.95 mm long, gradually long-acuminate, **dorsal leaves** somewhat asymmetrical, ovate, 1–1.3 mm long, gradually acuminate, concave, **lateral leaves** strongly asymmetrical, often cultriform, ovate-oblong, 0.85–1.2 mm long, ± abruptly acuminate, concave; **stem leaves** with margins distantly subserrulate throughout, plane;

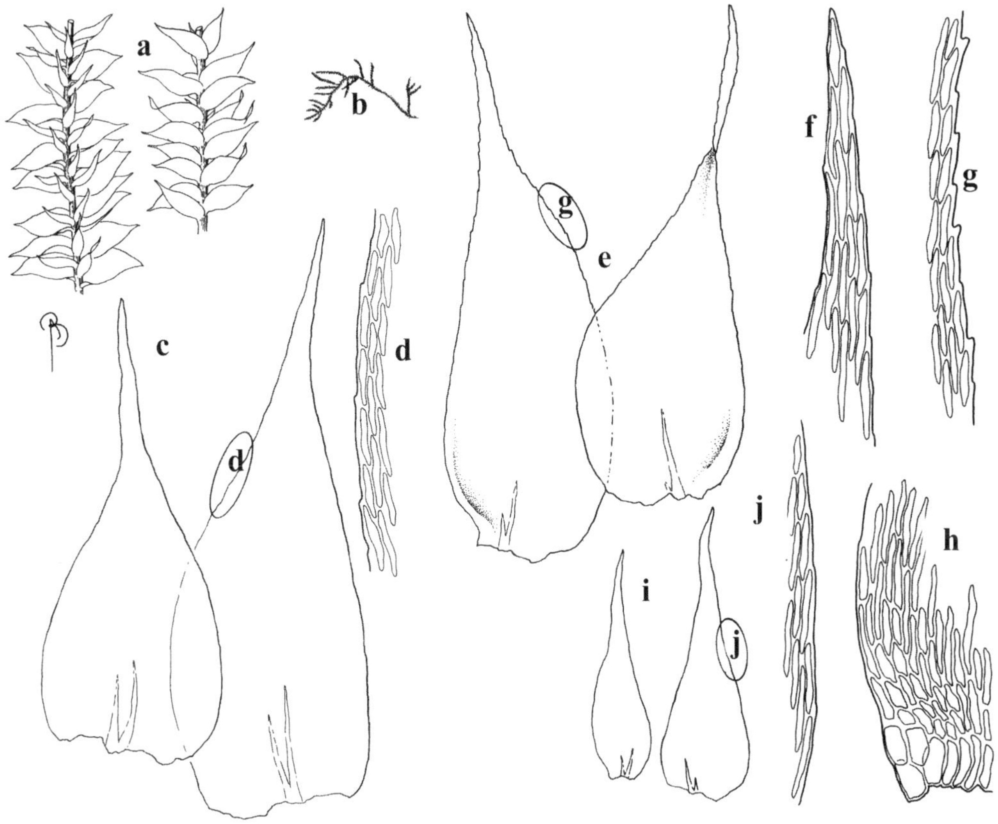

FIG. 30. *Rhacopilopsis trinitensis*. **a.** Portions of stems, ventral and dorsal views. **b.** Habit. **c.** Lateral stem leaves. **d.** Stem leaf margin. **e.** Lateral branch leaves. **f.** Branch leaf apex. **g.** Branch leaf margin. **h.** Basal marginal branch leaf cells. **i.** Ventral branch leaves. **j.** Ventral branch leaf margin. Reproduced with permission from Mem. New York Bot. Gard. 82: plate 121. figs. 1–10. 1998.

cells long-hexagonal to linear, 8–20:1, smooth but with thickened end walls, somewhat porose toward the insertion; alar cells mostly yellow, 2–4 cells enlarged in extreme basal angles, oblong, sometimes absent on one side of the lateral leaves. **Branch leaves with dorsal and lateral leaves** differentiated, **dorsal leaves** symmetrical, erect-spreading, lanceolate, 0.8–1.1 mm long, gradually acuminate, **lateral leaves** somewhat asymmetrical, wide-spreading, sometimes subfalcate, oblong–ovate, 1.1–1.3 mm long, gradually or ± abruptly acuminate, somewhat concave; **branch leaves** with margins subentire to serrulate throughout, the lateral leaves more strongly toothed, plane to narrowly recurved above; cells linear to linear-flexuose, 10–20:1, smooth, sometimes becoming shorter in the acumen, somewhat porose toward the insertion; alar cells yellow, 3–5 cells enlarged in extreme basal angles, oblong. **Sporophytes** not seen.

Habitat.—Moist lowland to lower montane forest, 200–1250 m, on tree trunks, especially at the bases, occasionally on logs.

VESICULARIA (Müll. Hal.) Müll. Hal.

Plants mostly medium-sized, in dark- to pale-green to golden, often extensive, thin mats. **Stems** creeping, irregularly to regularly pinnately branched; branches mostly simple, often straight when dry, somewhat recurved when moist. **Stem and branch leaves** differentiated, often contorted when dry, sometimes falcate-secund. **Lateral and ventral stem leaves** similar, **dorsal leaves** somewhat shorter pointed and with laxer

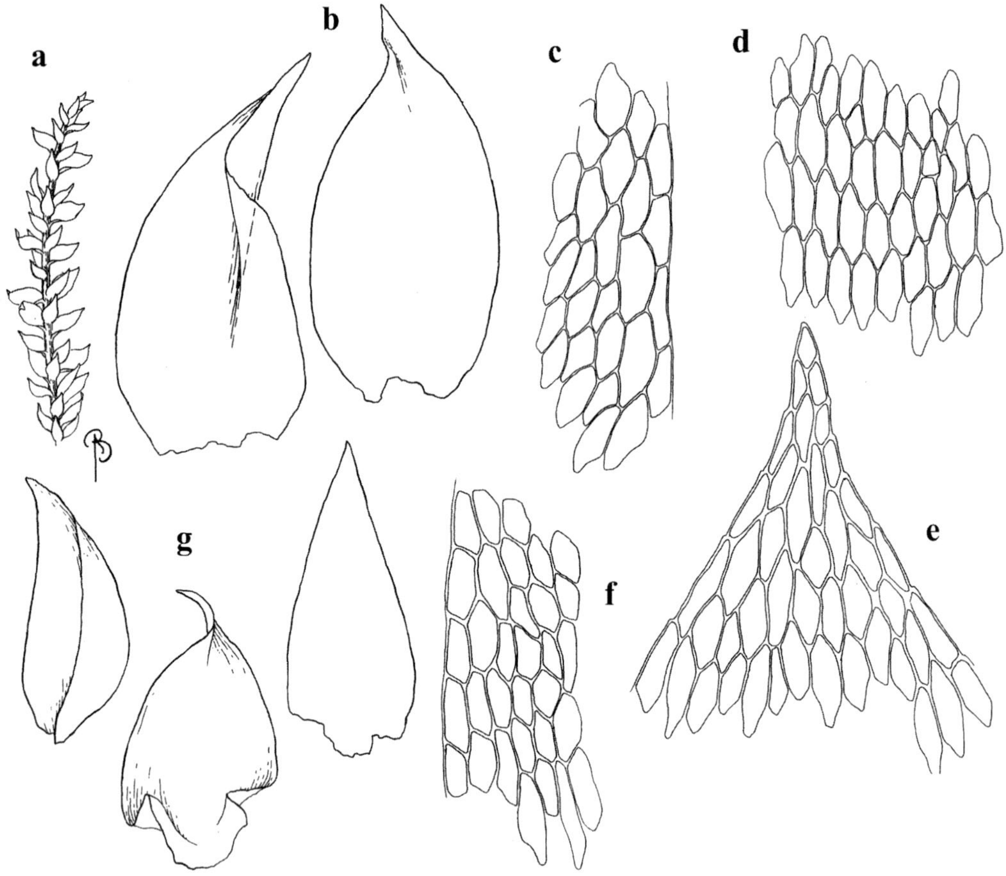

Fig. 31a. *Vesicularia vesicularis* var. *vesicularis*. **a.** Portion of stem. **b.** Stem Leaves. **c.** Stem leaf margin. **d.** Laminal cells, midleaf. **e.** Stem leaf apex. **f.** Branch leaves. **g.** Branch leaf margin. Reproduced with permission from Mem. New York Bot. Gard. 82: plate 122. figs. 6–10. 1998.

areolation, ovate-lanceolate to broadly ovate, acute to acuminate; margins entire to serrate, plane; **costa** short and double or absent; cells hexagonal to sublinear, broad, smooth, mostly thin-walled, becoming shorter and rectangular toward the insertion; alar cells not differentiated. **Branch leaves** often with differentiated **ventral leaves; dorsal and lateral leaves** lanceolate to ovate or broadly oblong-lanceolate, short- to long-acuminate; margins entire to serrulate, plane; **costa** short and double or absent; **ventral leaves** mostly lanceolate, short- to long acuminate; margins serrulate, plane; **costa** often lacking; cells longer than in lateral and dorsal leaves. **Setae** elongate, reddish, flexuose, curved just below the urn; **capsules** horizontal to pendent, short, ovoid to short-cylindrical, sometimes constricted below the mouth when dry; annulus differentiated; opercula apiculate to short conical-rostrate; peristome double, exostome teeth shouldered, bordered, on the front surface cross-striolate below, coarsely papillose above. **Calyptrae** cucullate, naked, smooth. **Spores** spherical, finely papillose, small.

31. Vesicularia vesicularis (Schwägr.) Broth. (**Figs. 31a, b**)

Plants small to medium-sized. **Stems** often fragile. **Stem leaves** broadly inserted, triangular, 0.8–1.2 mm long, gradually acuminate; margins entire to serrulate, plane; cells long-hexagonal, broad, 4–6:1, becoming

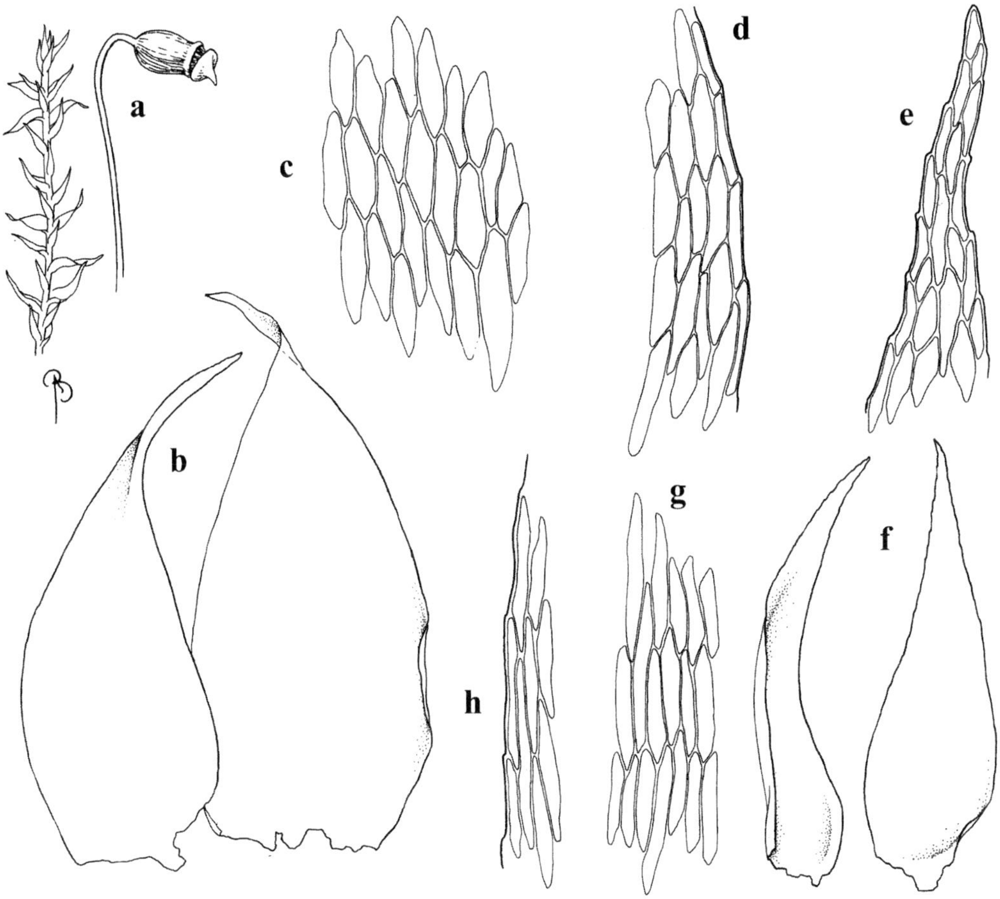

FIG. 31b. *Vesicularia vesicularis var. rutilans.* **a.** Portion of stem and sporophyte. **b.** Lateral leaves. **c.** Laminal cells, midleaf. **d.** Leaf margin. **e.** Leaf apex. **f.** Dorsal leaves. **g.** Laminal cells, midleaf. **h.** Dorsal leaf margin. Reproduced with permission from Mem. New York Bot. Gard. 82: plate 123. figs. 1–8. 1998.

broader and rectangular toward the insertion. **Dorsal and lateral branch leaves** ovate to broadly oblong-lanceolate, 0.7–0.95 mm long, apiculate to gradually acuminate; margins entire to serrulate, plane; cells lax, hexagonal to long-hexagonal, 1–3.5:1, smooth, becoming shorter to rectangular toward the insertion; **ventral leaves** sometimes differentiated, if so, then narrower, longer acuminate, and with narrower, longer cells, 1–8:1. **Setae** 1.2–2 cm long; **capsules** symmetrical to asymmetrical, ca. 1 mm long. **Spores** 11–14 µm diameter.

Notes.—*Vesicularia vesicularis* is extremely variable with several varieties recognized, although only two, var. *vesicularis* and var. *rutilans* (Bridel) W.R. Buck, are known from the study area. Buck (1998) recognizes both of the above varieties, however we believe they may be no more than ecological modifications to different habitats.

Habitat.—Wet lowland to lower montane forest, from near sea level to 2000 m, on a wide range of moist substrates from limestone and mud to logs.

1. Ventral leaves broad, with cells 1–2:1 ________ var. **vesicularis**
1. Ventral leaves narrow, with cells mostly 4–8:1 ________ var. **rutilans**

LEMBOPHYLLACEAE

Plants medium-sized to robust, in ± lustrous, pale-green to golden, sometimes black-tinged, loose to dense colonies. **Stems** creeping or pendent, occasionally erect, irregularly to subpinnately branched, the branches simple or often seemingly indeterminate and becoming stems. **Stem and branch leaves** not much differentiated or rarely differentiated (*Squamidium*), little altered when moist, mostly oblong-ovate, but also oblong-obovate or ± broadly rounded, sometimes panduriform, obtuse to hairpointed, concave, often ± clasping the stem, cordate to auriculate, sometimes decurrent; margins subentire to serrulate above, entire below, mostly broadly incurved, sometimes plane; **costa** mostly none or short and double, rarely single and slender (*Squamidium*); cells long-hexagonal to linear, smooth, mostly porose; alar cells mostly strongly differentiated, often in ± excavate groups, subquadrate to oblate, hyaline to colored, rarely few. **Setae** typically elongate, rarely very short, reddish or yellowish, often ± curved; **capsules** immersed to long-exerted, erect, cylindrical; annulus differentiated; opercula conical-rostrate, peristome double, smooth to papillose above, sometimes papillose throughout. **Calyptrae** cucullate or rarely mitrate (*Squamidium*), naked or hairy, smooth or roughened. **Spores** spherical, usually papillose, small to ± large.

SQUAMIDIUM (Müll. Hal.) Broth.

Plants medium-sized to ± robust, in lustrous, pale-green to golden, often black-tinged colonies. **Stems** creeping or pendent, irregularly branched, the branches mostly short and simple. **Stem and branch leaves** often differentiated, especially in shape and length of apex; margins subentire to serrulate or the stoloniferous stem leaves sometimes with large recurved teeth, plane or more typically incurved above; **costa** single, slender, often ending near base of the acumen; cells linear, smooth, mostly porose; alar cells strongly differentiated in discrete, ± excavate groups, subquadrate and thick-walled or less often short-rectangular and thin-walled. **Setae** short to very short from a hairy vaginula, yellow to orange; **capsules** immersed to short-exerted, erect, cylindrical. **Calyptra** mitrate, lobed at base, sparsely to densely erect-hairy, smooth. **Spores** lightly roughened.

Notes.—Some studies (Buck 1994; Ignatov & Hunnunen 2002) suggest that *Squamidium* might be better placed within the Brachytheciaceae.

32. Squamidium nigricans (Hook.) Broth. (**Fig. 32**)

Plants medium-sized, in mostly soft, bright- to yellow-green, often black-tinged colonies. **Stems** creeping to pendent, to 45 cm long, but mostly ca. 15–20 cm, slender, freely branched, the branches laxly to densely spaced, simple, 0.6–2 cm long. **Stem and branch leaves** differentiated, **stem leaves** clasping the stems, erect-spreading with spreading to squarrose long-apiculate apices, shortly and broadly oblong-ovate to deltoid, 1–2 mm long, **branch leaves** erect-spreading, broadly oblong-ovate, 0.6–1.7 mm long, abruptly apiculate to short-acute, all leaves concave, long-decurrent, weakly to not cordate at base; margins subentire to serrulate, mostly broadly incurved above; **costa** single, slender, sometimes very faint, ending above midleaf; cells linear-flexuose, ca. 15:1, smooth, thick-walled, scarcely porose; alar cells subquadrate or less often short-rectangular, firm-walled, in ± excavate groups but not inflated, extending up the margin in 6–10 rows. **Setae** very short, reddish, 2–3 mm long; **capsules** immersed, erect, cylindrical, 2–2.5 mm long; annulus of 2–3 rows of small thick-walled cells; opercula short-rostrate, 1.5 mm long; exostome teeth on the front surface ± smooth below, coarsely papillose above. **Calyptrae** sparsely erect hairs, smooth. **Spores** finely papillose, 17–23 μm diameter.

Habitat.—Primary to secondary lowland to montane forests, 0–3800 m, mostly epiphytic on trunks and branches, less often on logs, soil, and rocks.

LEUCOBRYACEAE

Plants in pale, greenish, whitish, or brownish tufts. **Leaves** crowded, erect to squarrose-recurved, sometimes flexuose or incurved to falcate-secund, thick and fleshy, lanceolate to ligulate from a somewhat broader

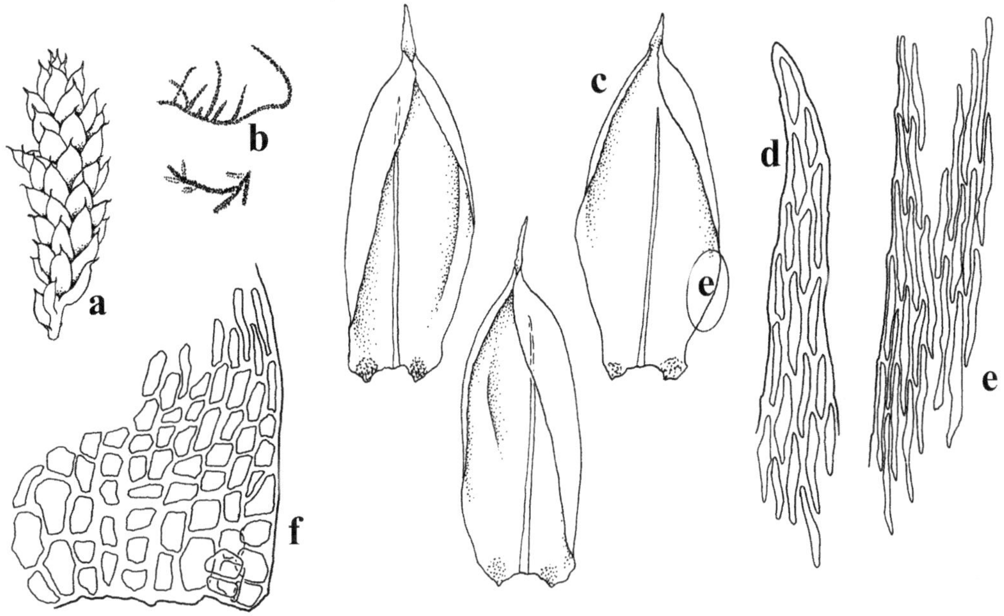

Fig. 32. *Squamidium nigricans*. **a.** Branch. **b.** Habit. **c.** Branch leaves. **d.** Leaf apex. **e.** Laminal cells at midleaf margin. **f.** Alar cells. Reproduced with permission from Mem. New York Bot. Gard. 82: plate 88, figs. 7–12. 1998.

base; **costa** filling the leaf apex and most of the base, consisting of a ± median layer of small chlorocysts enclosed on both surfaces by 1 or more layers of large, empty leucocysts; chlorocysts in surface view loosely reticulate in arrangement and rectangular or bead-like, in section 3–4-angled; leucocysts rectangular, with one or more large pores on connecting walls, in section not or slightly bulging; lamina constricted to the leaf base and consisting of 1–20 rows of hyaline cells on either side of the costa. **Setae** single, generally straight; **capsules** immersed or exerted, usually ribbed when dry, strumose or not at base; annulus usually none; opercula short- to long-rostrate; peristome single (or absent), of 8 or 16 lanceolate, bifid teeth, vertically striolate, papillose or smooth. **Calyptrae** mitrate or cucullate, naked, rarely ciliate-fringed at base. **Spores** lightly papillose.

1. Leaf blades ligulate; peristome teeth 8 or 16 (and ± paired), entire, smooth; in cross section chlorocysts in upper parts of the leaf triangular ______ **Octoblepharum**
1. Leaf blades lanceolate; peristome teeth 16, forked, vertically pitted-striolate; in cross section chlorocysts in upper parts of the leaf quadrangular ______ **Leucobryum**

LEUCOBRYUM Hampe

Plants in compact or loose cushions or mats; **stems** simple or forked to irregularly branched. **Leaves** appressed, imbricate, elliptic-ovate at base, erect-spreading, straight or flexuose, subtubulose, at times falcate-secund above, consisting mostly of a broad **costa**; apex acute or obtuse and cucullate; margins incurved above, entire or weakly denticulate at the apex; cross-section at base with a median row of small, green quadrangular cells (chlorocysts) enclosed by 2–8 layers of enlarged, hyaline cells; leaf lamina present from leaf base to midleaf, consisting of a narrow (2–12 rows of cells) marginal band of hyaline, linear, long-rectangular, short-rectangular, or quadrate cells. **Setae** elongate; **capsules** cylindrical, curved or rarely erect, commonly strumose, ribbed or smooth when dry; annuli mostly persistent; peristome teeth divided 1/2 of the distance to the base, vertically-striate, papillose or smooth on outer surface.

1. Plants medium-sized, 10–30 mm high, often reddish tinged; leaves falcate-secund, narrowly acute, 3–7 mm long; upper chlorocysts rectangular in surface view ______ **Leucobryum martianum**
1. Plants small-sized, 5–7 mm high, never reddish tinged; leaves straight, broadly acute to obtuse, 2–3 mm long; upper chlorocysts subdivided and forming chains of rounded to quadrate cells in surface view ______ **Leucobryum subobtusifolium**

33. Leucobryum martianum (Hornsch.) Hampe (**Fig. 33**)

Plants glaucous green, 1–3 cm high. **Leaves** 3–7 mm long, commonly reddish tinged, particularly at the base, crowded and falcate-secund or occasionally erect-spreading, long-lanceolate throughout or gradually narrowed from an ovate-lanceolate base to a slender subtubulose acute apex; **costa** in cross section at base consisting of a median layer of chlorocysts between a single layer of dorsal and ventral leucocysts throughout the leaf; leaf lamina narrow, consisting of 3–7 rows of long–rectangular, firm-walled cells at midleaf and 2–5 rows of short-rectangular cells at base. **Setae** red, straight or flexuose, 10–15 mm long; **capsules** inclined to curved, 1 mm long, furrowed when dry, strumose; peristome teeth red. **Spores** 15–22 µm, smooth.

Habitat.—Lowland and savanna forest, 0–2400 m, on logs, tree trunks, roots of epiphytic and terrestrial plants, and terrestrial. Widespread in the Neotropics.

34. Leucobryum subobtusifolium (Broth.) B.H. Allen (**Fig. 63 a, b**)

Plants small, glaucous green, 3–7 mm high. **Leaves** 2–3 mm long, erect at base, weakly flexuose above, lanceolate to narrowly ovate-lanceolate, gradually narrowed from base to apex; apex broadly acute to obtuse and apiculate; margins erect to incurved, entire; **costa** in cross-section throughout the leaf with a single layer of dorsal leucocysts, a median layer of chlorocysts and a single layer of ventral leucocysts, in surface view the dorsal leucocysts enlarged, rectangular, not subdivided; upper chlorocysts subdivided and forming chains of rounded to quadrate cells; leaf lamina poorly developed, 1–2 rows of linear cells above, 2–4 rows of rectangular cells at base. **Sporophytes** unknown.

Habitat.—Wet to semi-wet lowland forest, 70–700 m, on logs, base and trunk of trees. The single collection we obtained came from the base of *Mauritia flexuosa* in a palm swamp.

OCTOBLEPHARUM Hedw.

Plants glaucous green, yellow, reddish or purple, dull to shiny, occasionally with a metallic luster; **stems** erect, sparingly branched. **Leaves** erect-patent, flexuose or reflexed above midleaf, ligulate, obtuse-apiculate to shortly acuminate, flattened or terete-triangular; margins entire to dentate, occasionally undulate at apex, border present or absent; basal lamina with porate leucocysts in 1(–2) layers; median cross-section of leaf with 3–4(–5) layers of leucocysts surrounding a central layer of mostly triangular to tear-drop-shaped chlorocysts. **Setae** yellowish to dark orange-red, short or long; **capsules** oblong or cylindrical, orange to dark red; peristome teeth 8–16, smooth to vertically striate-reticulate or with semi-circular or bar-like thickenings; opercula rostrate. **Calyptrae** cucullate, naked, and smooth. **Spores** yellow-green to brownish-green.

1. Leaves fragile, often broken, yellow-green, purple or light green above, purple below; peristome teeth 16 ______ **Octoblepharum pulvinatum**
1. Leaves sturdy, white-green to yellow-green or red to pink at base; peristome teeth 8.
    2. Ventral laminal cells long hexagonal to rectangular; leaf apex dentate; capsules oval; setae 4–8 mm long; peristome teeth broadly triangular with faint, delicate trabeculae ______ **Octoblepharum albidum**
    2. Ventral laminal cells quadrate; leaf apex more or less entire; capsules cylindrical; setae greater than 10 m long; peristome teeth lanceolate with thick, strongly projecting trabeculae ______ **Octoblepharum cylindricum**

35. Octoblepharum albidum Hedw. (**Fig. 34**)

Plants in dense to loose turfs and cushions, glaucous green to dull white, often with a pink to purple tone at base, 0.5–3 cm high. **Leaves** erect-patent to curved spreading, flattened, strap-shaped, apiculate, 3–8 mm long, margins with inconspicuous border, 1–2 cells wide, sometimes undulate due to the presence of marginal inflated leucocysts in rows of 3–5 (or more); basal lamina with elongate hexagonal to rectangular

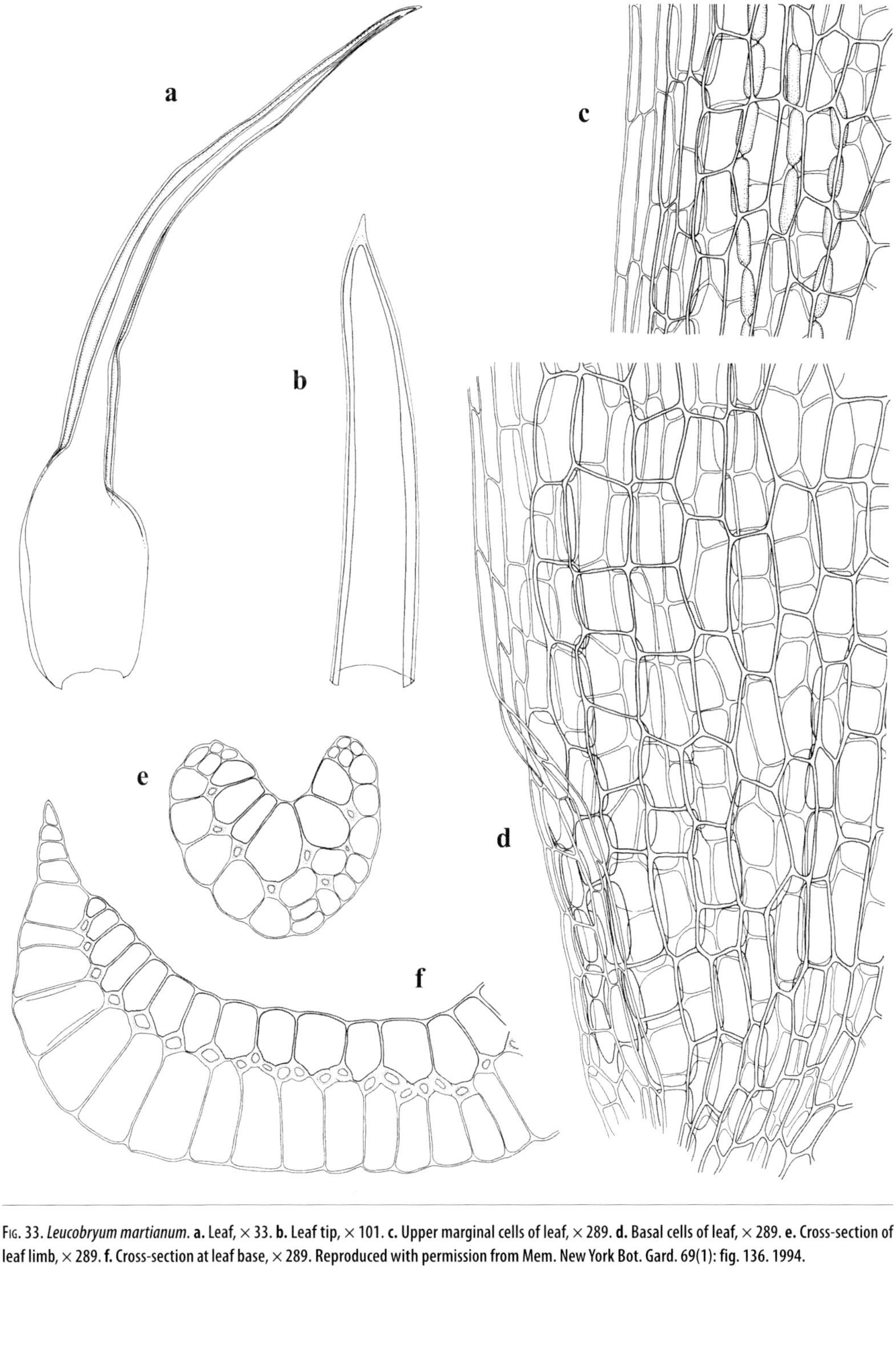

FIG. 33. *Leucobryum martianum*. **a.** Leaf, × 33. **b.** Leaf tip, × 101. **c.** Upper marginal cells of leaf, × 289. **d.** Basal cells of leaf, × 289. **e.** Cross-section of leaf limb, × 289. **f.** Cross-section at leaf base, × 289. Reproduced with permission from Mem. New York Bot. Gard. 69(1): fig. 136. 1994.

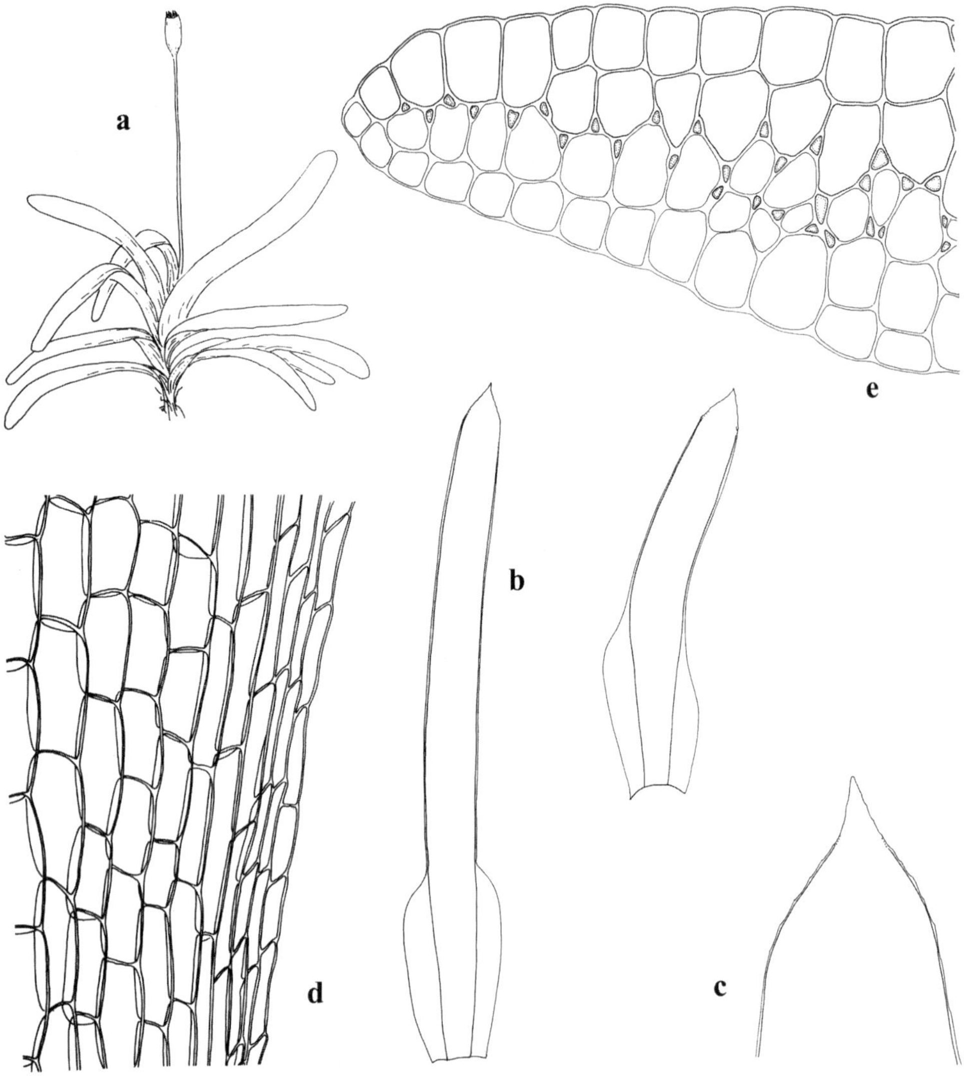

FIG. 34. *Octoblepharum albidum*. **a.** Habit, × 4. **b.** Leaves, × 33. **c.** Leaf tip, × 101. **d.** Basal cells of leaf, × 289. **e.** Cross-section of leaf limb, × 289. Reproduced with permission from Mem. New York Bot. Gard. 69(1): fig. 137. 1994.

leucocysts, in 1(–2) layers, pores narrow and inconspicuous; cross-section of leaf blade at middle elliptic with a central layer of chlorocysts surrounded by 3–4 layers of leucocysts on each surface. **Setae** 2–8 mm long, yellow to orange-red; **capsules** erect to subinclined, oval, orange-red, darker and contracted at mouth; peristome teeth 8, smooth to vertically striate or reticulate. **Spores** 20–25 µm in diameter.

Habitat.—Lowland to lower montane forest, 20–1675 m, on bark at base of trees (particularly palms) and shrubs, logs, soil and rocks.

36. Octoblepharum cylindricum Schimp. ex Mont. (Fig. 64)

Plants in dense tufts, glaucous green, pinkish towards base, 0.5–2 cm high. **Leaves** erect to curved outwards above middle, strap-shaped, apiculate to shortly acuminate, not fragile, 4–8 mm long; margins conspicu-

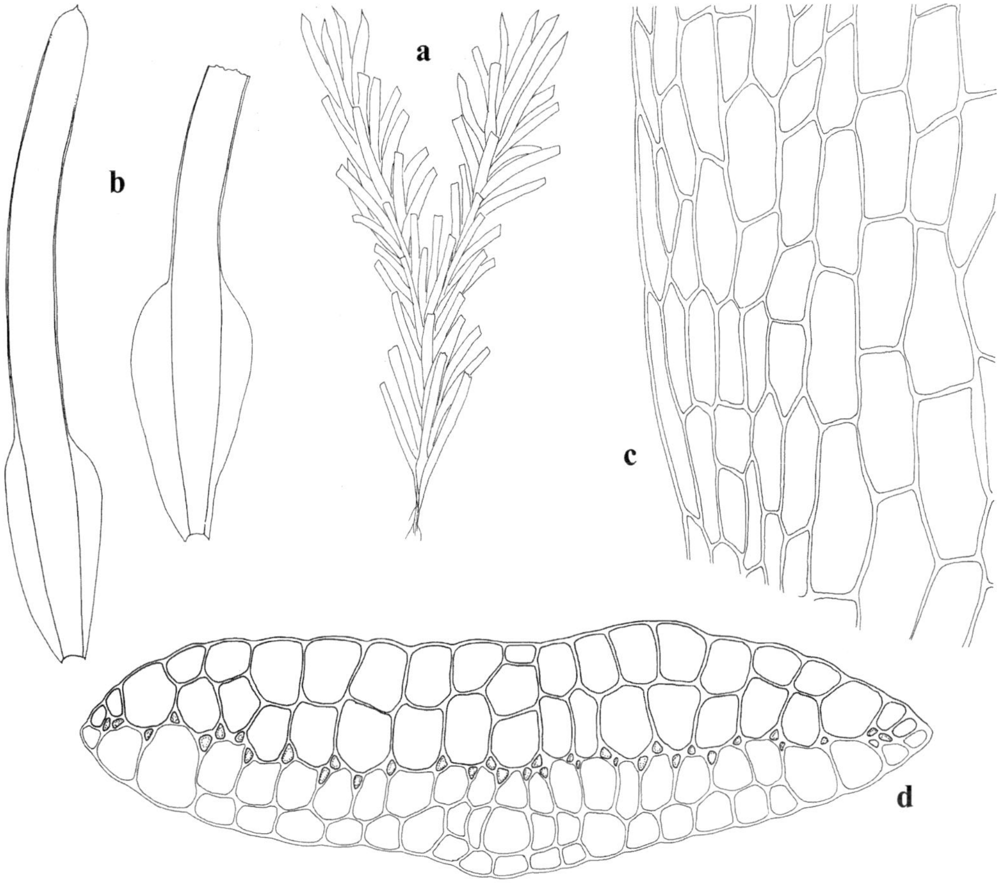

FIG. 35. *Octoblepharum pulvinatum*. **a.** Habit, × 2. **b.** Leaves, × 33. **c.** Basal cells of leaf, × 289. **d.** Leaf section, × 289. Reproduced with permission from Mem. New York Bot. Gard. 69(1): fig. 139. 1994.

ous at apex, 1–2 cells wide, entire to undulate towards apex, rarely toothed; basal hyaline lamina gradually tapering to the costal area with quadrate leucocysts at center next to costa, short hexagonal cells near border, quadrate to rectangular at base; cross-section of leaf elliptic, with a central layer of chlorocysts surrounded by 2–3 layers of leucocysts. **Setae** smooth, 8–20 mm long, straw-colored to light orange-red; **capsules** long cylindrical, erect to curved or subinclined, urn slightly contracted below mouth, peristome teeth 8, with thickened midline and trabeculae, and diagonal to vertical striations. **Spores** 8–21 µm, strongly granular.

Notes.—*Octoblepharum cylindricum* differs from *O. albidum* by its longer setae, cylindrical capsules, entire leaf apices, and strongly trabeculate peristome teeth.

Habitat.—Lowlands to submontane forest, 200–2180 m, in dense turfs on tree trunks, moist sand and stones.

37. Octoblepharum pulvinatum (Dozy & Molk.) Mitt. (Fig. 35, 65)

Plants in loose to dense turfs, glaucous green, occasionally with a tinge of pink, 0.8–2 cm high; stems with erect, more or less parallel innovations. **Leaves** fragile or, if sturdy, some broken, erect-spreading to flexuose-secund, shiny with a broad hyaline lamina abruptly narrowed to the strap-shaped upper leaf, mucronate to nearly truncate, 5–9 mm long; margins entire, sharply to inconspicuously and obtusely dentate at apex, smooth below, slightly undulate above; upper laminal cells irregularly short hexagonal, quadrate or pentagonal, 45–60 µm in longest dimension; leaf cross-section elliptic at middle, central layer of chlorocysts surrounded by 2(–3) layers of leucocysts on each surface. **Setae** 5–8 mm long, light yellow above, orange-red below; **capsule** erect to subinclined, cylindrical, orange-red, slightly contracted under mouth, bulging papillose at base; peristome of 16 elongate, paired teeth, articulate, trabeculate with papillose, bar-like thickenings and central circular depressions. **Spores** 15–18 µm in diameter.

Habitat.—Lowland to lower montane forests, 30–2100 m, epiphytic on trees (particularly palms), shrubs, logs, humus, and occasionally on rock.

METEORIACEAE

Plants slender to robust, in dull or less often lustrous, soft to stiff, green to golden, frequently black-tinged, typically pendent colonies; usually epiphytic. **Stems** elongate, at least initially creeping along substrate, but then becoming ascendent to pendent; primary and secondary stems not differentiated. **Stem and branch leaves** mostly not differentiated except in size, appressed to spreading, lanceolate to ovate, mostly acuminate, sometimes hairpointed, often concave, sometimes plicate, often with a ± clasping base and either cordate or ± auriculate; margins mostly serrulate almost throughout, plane to incurved or recurved; **costa** single or less often short and double or absent; cells short to linear, smooth or more often papillose on both surfaces, either with a single papilla over the lumina or pluripapillose over the lumina or the walls or both, sometimes seriately arranged; alar cells scarcely differentiated or differentiated in small areas, rarely numerous. **Setae** short to elongate; **capsules** immersed to long-exerted, erect to suberect, globose to cylindrical; annulus mostly differentiated; opercula typically obliquely conical-rostrate; peristome double, often inserted somewhat below the mouth. **Calyptrae** mitrate or cucullate, naked or hairy, smooth or roughened. **Spores** spherical, variously papillose.

ZELOMETEORIUM Manuel

Plants medium-sized, in green to golden pendent masses. **Stems** pendent to creeping, irregularly to pinnately branched; flagellate branches sometimes present. **Stem and branch leaves** differentiated, **stem leaves** ± auriculate and strongly clasping the stem, longer and more slenderly acuminate, **branch leaves** spreading to squarrose-recurved, not clasping at base, lanceolate-ovate to suborbicular, apiculate to piliferous; margins subentire to sharply serrulate, plane throughout or revolute below; **costa** single, slender, ending near or somewhat above midleaf; cells long-hexagonal to linear-flexuose, smooth, sometimes ± porose; alar cells somewhat to well-differentiated, subquadrate to short-rectangular. **Setae** relatively short, reddish; **capsules** short-exerted, erect, long-cylindrical; annulus differentiated; opercula conical-rostrate; peristome pale, papillose on the front surface. **Calyptrae** mitrate, lobed at base, erect-pilose, roughened above. **Spores** spherical, smooth, or roughened.

Notes.—Some studies (Buck 1994; Ignatov & Hunnunen 2002) suggest that *Zelometeorium* might be better placed within the Brachytheciaceae.

38. Zelometeorium patulum (Hedw.) Manuel (Fig. 36)

Plants often in extensive, green to golden, pendent festoons. **Stems** pendent to creeping, to ca. 30 cm long, the branches short, mostly less than 1 cm long and the branch apices rarely becoming flagellate. **Stem leaves** erect to erect-spreading from a clasping base, narrowly to broadly ovate, ca. 2 mm long, abruptly acuminate, the acumen flexuose and often piliferous, broadly auriculate, the auricles ca. 30–40% of the leaf base (excluding the acumen); margins subentire to serrulate throughout; **costa** single, ending near midleaf; cells linear, smooth, usually not or scarcely porose, but becoming obviously so toward the insertion; alar cells scarcely differentiated to fairly obvious in small, excavate groups, subquadrate to short-rectangular, somewhat porose. **Branch leaves** densely disposed, wide-spreading to squarrose-recurved, sometimes somewhat contorted when dry, mostly ovate, 0.85–2.2 mm long, short- to long-acuminate, sometimes

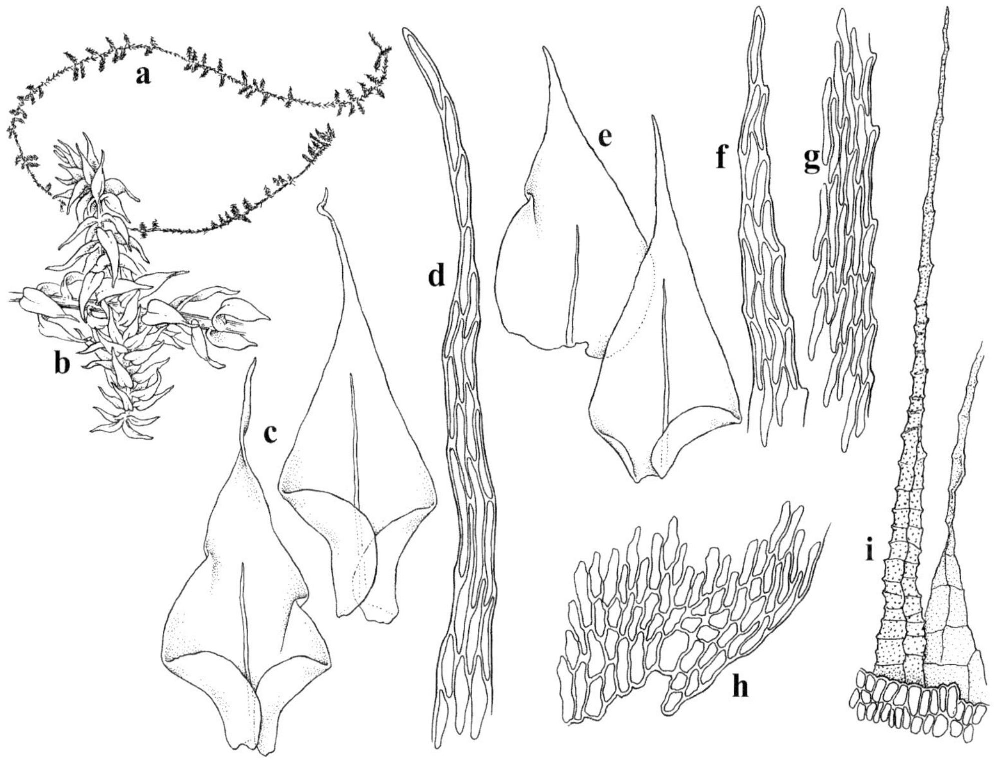

FIG. 36. *Zelometeorium patulum*. **a.** Habit. **b.** Portion of stem with branches. **c.** Stem leaves. **d.** Stem leaf apex. **e.** Branch leaves. **f.** Branch leaf apex. **g.** Branch leaf margin. **h.** Basal marginal stem leaf cells. **i.** Portion of peristome. Reproduced with permission from Mem. New York Bot. Gard. 82: plate 99, figs. 10–18. 1998.

piliferous, the base mostly cordate, not clasping; margins subentire, denticulate or serrulate throughout; **costa** ending near midleaf or to ca. 3/4 the leaf length; cells fusiform to linear-flexuose, smooth, sometimes ± porose, becoming rectangular and porose toward the insertion; alar cells indistinct to obvious in small, well-marked groups, subquadrate, sometimes slightly porose. **Setae** ca. 3 mm long, straight or curved; **capsules** short-exerted, erect, long-cylindrical, 2–4 mm long; peristome yellow, scarcely bordered, sparsely papillose below, densely papillose-spiculose above. **Calyptrae** erect-pilose, the hairs 1–3(4) seriate. **Spores** low papillose, 14–23 µm diameter.

Habitat.—Humid lowland to montane forest, 20–3200 m, on trunk and branches of treelets and trees, also palms, occasionally epiphyllous or over rocks, humus, and leaf litter.

NECKERACEAE

Plants medium-sized to rather large and robust, forming tufts or mats, light to dark green, yellowish-green to golden-brown. **Primary stems** short to long creeping, leaves scale-like or eroded; radiculose beneath stems. **Secondary stems** ascending to pendent or stiffly erect and dendroid or subfrondose, stipitate or not, when stipitate leaves triangular to ovate-triangular; irregularly to regularly pinnately branched, distal branch flagelliform or not. **Leaves** crowded to somewhat distant, complanate or radially foliate, ovate- to oblong-lanceolate, oblong-lingulate, or cultriform, smooth, plicate or undulate, symmetrical to strongly asymmetrical, apex obtuse to acute, truncate or acuminate, base not or short decurrent on one side; margins plane to folded at base (usually along one side), entire or distally dentate, serrulate to serrate; **costa** single and

1/2–4/5 lamina length or short to somewhat long forked, occasionally absent. **Setae** very short to elongate, smooth or occasionally roughened above, reddish or less often yellow; **capsules** immersed to long-exerted, mostly cylindrical, erect and symmetrical; peristome double; endostome often somewhat reduced. **Calyptrae** cucullate, rarely mitrate-campanulate, naked or rarely sparsely hairy or ramentose.

NECKEROPSIS Reichardt

Plants small to robust, in mostly pale to yellowish green, often large, epiphytic colonies, not stipitate-frondose. **Primary stem** stoloniform, leaves remote, scale-like; **secondary stems** often flexuose, strongly complanate. **Secondary stem and branch leaves** not differing or only differing in size, apiculate, round or truncate at apex, asymmetrical, often transversely undulate; margins crenulate at apex; **costa** usually single and extending beyond midleaf, or short and forked. All median laminal cells usually pellucid, linear, oblong or rhomboidal, the walls incrassate; upper cells shorter and wider; basal cells linear, the walls thick and porose; alar cells not differentiated. **Setae** short, smooth; **capsules** immersed or exerted; annulus absent; opercula conical, with a small oblique beak; exostome teeth papillose; endostome basal membrane low, the segments linear, stout, perforate, as long as the exostome teeth. **Calyptrae** cucullate to rarely mitrate, with or without hairs. **Spores** finely papillose.

1. Leaves smooth or nearly so, flat when moist, not auriculate; calyptrae cucullate ______ **Neckeropsis disticha**
1. Leaves strongly undulate moist or dry, auriculate; calyptrae mitrate______ **Neckeropsis undulata**

39. Neckeropsis disticha (Hedw.) Kindb. (**Fig. 37**)

Plants small to medium-sized, in loose, pale- to yellow-green mats. **Stems** up to 10 cm long, irregularly and often sparsely branched; branches blunt or slightly tapered. **Leaves** crispate when dry, smooth or slightly undulate when moist, 1–1.5 × 0.5 mm, ligulate, moderately auriculate, broadly rounded-obtuse or rarely truncate; margins entire, except at the crenulate apex; **costa** ending 1/2–3/4 the leaf length; upper cells hexagonal, firm-walled, somewhat longer, rectangular, and porose toward the base; **capsules** up to 1.25 mm long, brown or straw-colored. **Calyptrae** cucullate. **Spores** 14–16 µm diameter.

Habitat.—Moist or wet lowland forests, 0–1120 m, on exposed roots, trunks, and branches of trees, occasionally on rocks. Very common in the study area.

40. Neckeropsis undulata (Hedw.) Reichardt (**Fig. 38**)

Plants small to medium-sized, in dense, pale- to yellow-green mats. **Stems** up to 7 cm long, freely and irregularly branched; branches blunt. **Leaves** strongly undulate, 1.4–2.5 × 0.5–0.8 mm long, oblong-ligulate, truncate to broadly rounded-obtuse; margins entire, except at the weakly crenulate apex; **costa** single, slender, ending ca. 3/4 the leaf length; upper cells rhombic, becoming longer-rectangular below the leaf middle; **capsules** up to 1.5 mm long, yellow-brown. **Calyptrae** mitrate. **Spores** 20–25 µm diameter.

Habitat.—Humid lowland to lower montane forest, 0–1900 m, on trunks, at bases and branches of trees, rarely on rocks. Very common in the study area.

ORTHOTRICHACEAE

Plants small to robust, in cushions, tufts or mats on rocks or trees. **Stems** erect, erect-ascending or creeping with erect-ascending branches, often densely or sparsely tomentose. **Leaves** crowded, erect to crispate or spirally twisted when dry, spreading to squarrose when wet, more or less keeled, lanceolate, ovate-lanceolate, oblong-lanceolate, ligulate or lingulate; margins plane or revolute, rarely incurved, generally entire to serrate or denticulate; **costa** single, strong, ending near the apex; upper cells generally small, irregularly rounded and thick-walled, papillose or smooth, basal leaf cells generally elongate, less frequently short and rounded, tuberculate or smooth; alar cells undifferentiated, weakly differentiated or rarely strongly differentiated. Brood bodies sometimes present in leaf axils, on rhizoids or leaves. **Setae** smooth or roughened; **capsules** immersed, emergent, or exserted, erect and symmetrical, smooth or ribbed; opercula rostellate or rostrate;

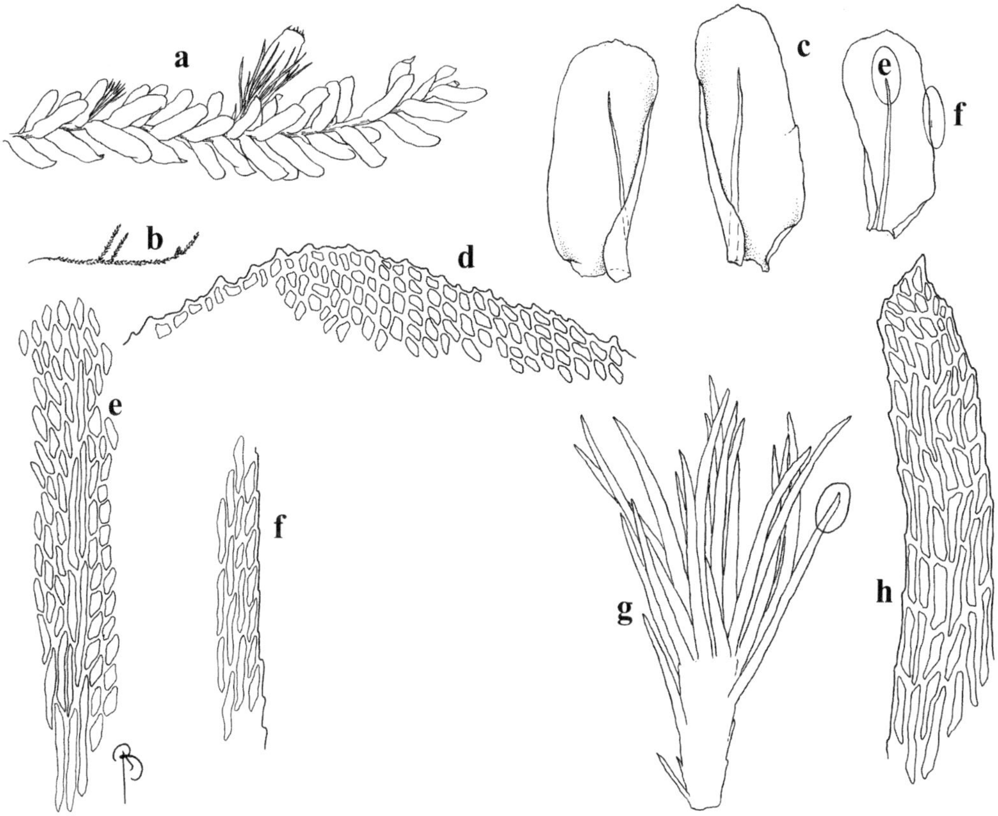

Fig. 37. ***Neckeropsis disticha*. a.** Portion of stem with sporophyte. **b.** Habit. **c.** Leaves. **d.** Leaf apex. **e.** Laminal cells and costa apex. **f.** Leaf margin. **g.** Vaginula with ramenta. **h.** Apex of ramentum. Reproduced with permission from Mem. New York Bot. Gard. 82: plate 40, figs. 9–16. 1998.

peristome double but often absent, exostome teeth 8 or 16. **Calyptrae** campanulate, conical or cucullate, naked or hairy, plicate or non-plicate. **Spores** isosporous or anisosporous, sometimes multicellular.

GROUTIELLA Steere in H.A. Crum & Steere

Plants slender to medium-sized, in dark-green, yellow-green, reddish-green, reddish-brown to olive-brown, tomentose mats or cushions. **Primary stems** creeping; **secondary stems** erect-ascending, irregularly branched. **Leaves** contorted, spirally twisted, crisped to rugose when dry, erect-spreading to wide-spreading when wet, keeled, linear-lanceolate, lanceolate, oblong- to ovate-lanceolate or lingulate; apices obtuse, acute, acuminate, to mucronate, rarely with fragile tips; margins entire above, entire to serrulate at base, bordered at base by several rows of narrow, elongate cells from 1/4 to 2/3 the leaf length; **costa** strong, subpercurrent or excurrent; cells incrassate, upper cells rounded-hexagonal, smooth to mammillose, basal cells rounded-rectangular, smooth or rarely with a few tubercula, those near the costa somewhat enlarged and thin-walled, at times with enlarged basal teeth at leaf insertion. **Setae** smooth; capsules oblong-cylindrical or obovoid, smooth or furrowed; opercula rostrate; peristome rudimentary, consisting of a low, papillose membrane. **Calyptrae** mitrate-campanulate, laciniate or lobed at base, plicate or smooth, naked or sparsely hairy. **Spores** isosporous or anisosporous.

41. Groutiella tomentosa (Hornsch.) Wijk & Margad. (**Fig. 39**)

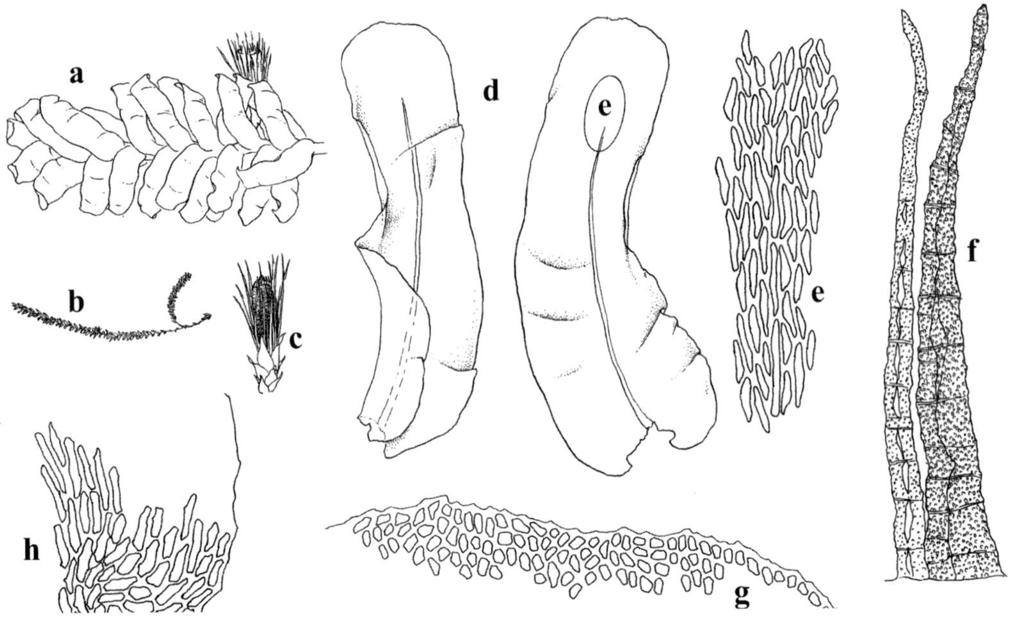

Fig. 38. ***Neckeropsis undulata*.** **a.** Portion of stem with sporophyte. **b.** Habit. **c.** Sporophyte. **d.** Leaves. **e.** Laminal cells. **f.** Portion of peristome. **g.** Leaf apex. **h.** Basal marginal leaf cells. Reproduced with permission from Mem. New York Bot. Gard. 82: plate 40, figs. 1–8. 1998.

Plants small to medium-sized, greenish yellow to reddish yellow above, brown below. **Stems** usually strongly creeping, branches to 10(–20) mm long, strongly tomentose. **Leaves** firm, erect below, appressed and tightly spirally twisted around the stem above when dry, erect-spreading and undulate when wet, 1.5–2 mm long; leaves dimorphic, lower leaves oblong-lanceolate, acute, non-fragile, upper leaves from a broadly oblong base abruptly constricted to a long, linear, fragile subula; margins entire above, plane below, recurved in lower 1/2, bordered in the lower 1/3–1/2 by 1–4 rows of elongate, thick-walled cells; **costa** subpercurrent to percurrent in lower leaves, percurrent and filling the subula in the upper leaves; upper cells 4–8 µm, rounded-hexagonal, not or weakly bulging mammillose, basal cells smooth or tuberculate, those near the costa somewhat enlarged and thin-walled, those near the insertion rounded-rectangular, basal marginal cells inflated more or less serrulate by projecting cell end walls, frequently with rhizoids. **Setae** 4–10 mm long, smooth; **capsules** dark-red, 2–3.5 mm long, oblong-cylindrical, smooth to weakly furrowed, wrinkled at neck; opercula erect-rostrate, 1.5–2 mm long; annulus non-revoluble, with fragments adhering to capsule mouth; peristome rudimentary, reduced to a short papillose membrane projecting 180 µm above the capsule mouth. **Calyptrae** strongly laciniate, plicate above, naked, 2–3 mm long. **Spores** isosporous, 18–28 µm diameter, firm-walled, smooth to weakly papillose.

Habitat.—Lowland to lower montane forests, 5–1880 m, on bark of tree trunks and branches, logs and rocks.

SCHLOTHEIMIA Brid.

Plants small to medium-sized, dark-green above, reddish-brown below, thickly matted with dense reddish tomentum. **Primary stems** creeping, with erect or ascending **secondary stems or branches**. **Leaves** imbricate, spirally twisted around stems when dry, erect to erect-spreading, apices recurved when wet, linear-lanceolate, lingulate to oblong-ovate; margins entire, not bordered; **costa** strong, percurrent to excurrent; upper cells smooth, incrassate, rounded to rhomboidal, basal cells elongate-rhomboidal, incrassate and porose, occasionally with projecting end walls. **Setae** erect, smooth; **capsules** usually exerted, elliptical to cylindrical, smooth or weakly furrowed; opercula short-rostrate; exostome teeth 16, densely horizontally

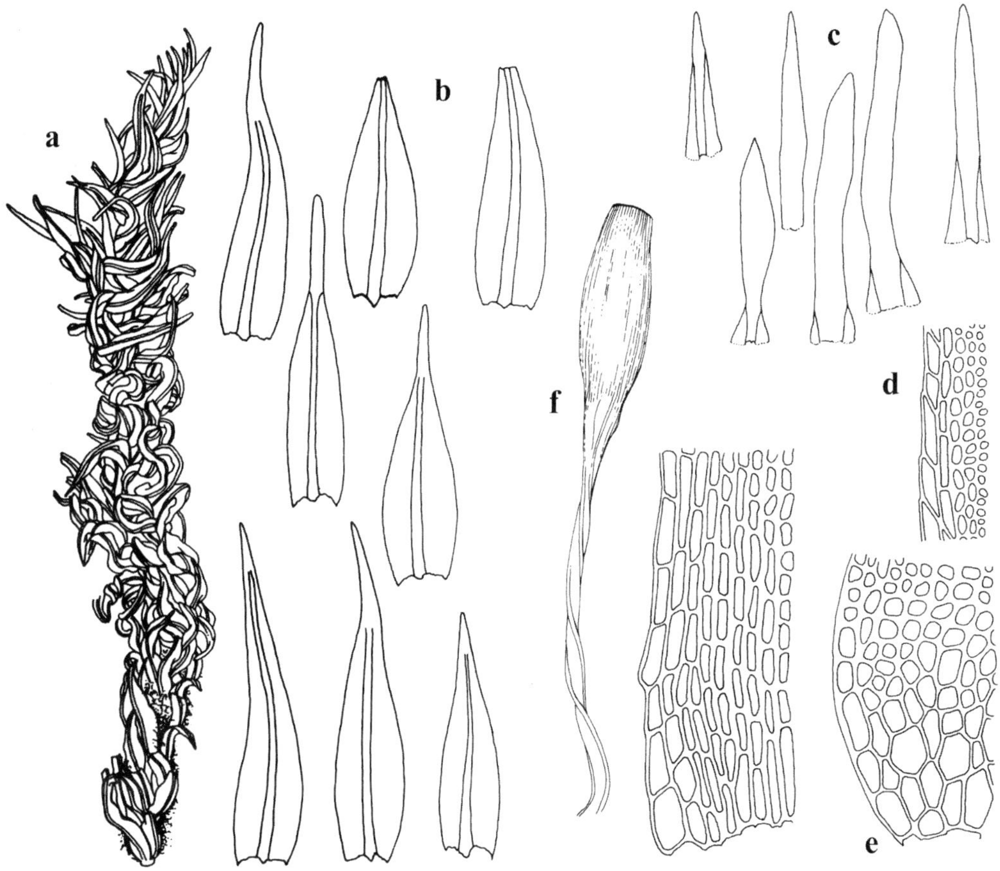

FIG. 39. *Groutiella tomentosa*. **a.** Branch, × 16. **b.** Leaves, × 85. **c.** Leaf tips, × 85. **d.** Leaf margin half way up, × 419. **e.** Basal marginal cells of leaf, 2 views, × 419. **f.** Capsule, × 16. Reproduced with permission from Mem. New York Bot. Gard. 69(2): fig. 481. 1994.

striate on outer surface. **Calyptrae** mitrate-campanulate, 4–6 lobed, smooth or papillose. **Spores** isosporous or anisosporous, papillose.

42. Schlotheimia rugifolia (Hook.) Schwägr. (Fig. 40)

Plants small to medium-sized. **Secondary stems** erect, 2–25 mm long, densely tomentose. **Leaves** rugose, erect-contorted to spirally twisted around stem when dry, erect-patent to patent, reflexed at tips when wet, 1.3–2 mm long, ovate-oblong to oblong-lingulate; apices apiculate to stoutly mucronate; margins entire, plane or weakly recurved below, plane above; **costa** excurrent, filling the apiculus; upper cells 6–12 µm, rounded-quadrate to oblate, smooth, incrassate, basal cells 16–32 µm long, long-rhomboidal, incrassate and porose with narrow lumina. Filiform, densely papillose brood bodies occasionally present on lower, dorsal costa surface. **Setae** 2–4 mm long; **capsules** 1.5–2 mm long, oblong-cylindrical, smooth, wrinkled or furrowed when dry; opercula to 1 mm long; exostome teeth red, with a narrow median furrow, reflexed when dry, endostome yellow. **Calyptrae** campanulate, papillose above, lobed at base, 3–3.5 mm long. **Spores** anisosporous, 14–26 µm diameter, lightly papillose and 30–50 µm diameter, densely papillose.

Habitat.—Humid lowland to lower montane forests, 10–2200 m, on bark of tree trunks and upper branches, fallen logs, and occasionally on boulders.

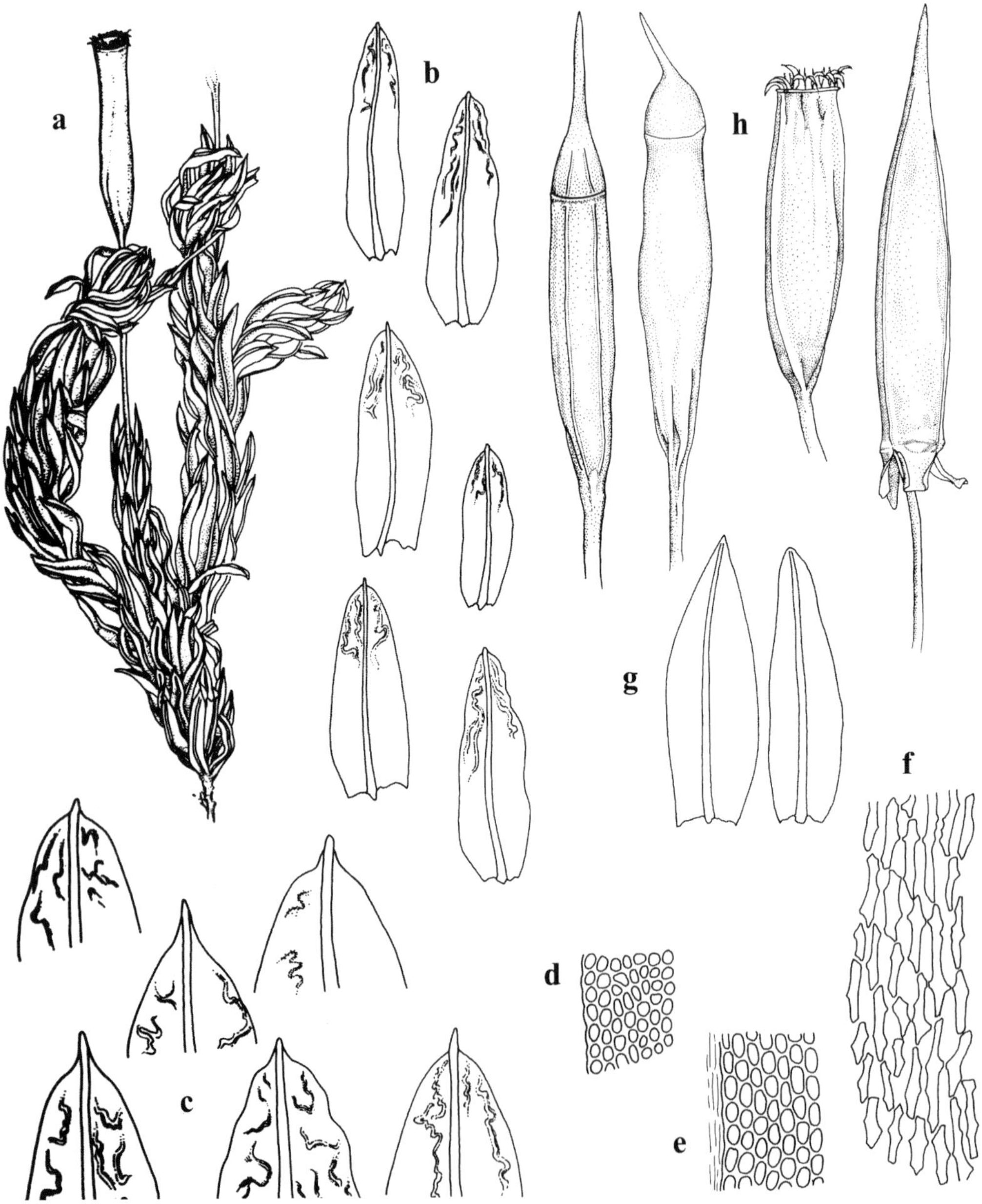

FIG. 40. *Schlothemia rugifolia.* **a.** Habit, × 11. **b.** Leaves, × 134. **c.** Leaf tips, × 134. **d.** Upper marginal cells of leaf, × 419. **e.** Upper cells of leaf near costa, × 419. **f.** Inner basal cells of leaf, × 419. **g.** Perichaetial leaves, × 134. **h.** Capsules and calyptra, × 16. Reproduced with permission from Mem. New York Bot. Gard. 69(2): fig. 490. 1994.

PHYLLOGONIACEAE

PHYLLOGONIUM Brid.

Plants medium-sized to robust, green to golden to bronze, pendent, often in extensive, epiphytic colonies. **Primary stems** creeping, tightly attached to the substrate, with reduced, scale-like leaves; **secondary stems** irregularly branched, pendent, strongly complanate-foliate. **Secondary stem and branch leaves**

not differentiated, conduplicate, cucullate, auriculate; margins subentire, incurved; **costa** short and double or absent; cells linear, smooth, thick-walled, porose; alar cells differentiated in small areas, thick-walled, porose. **Setae** short, smooth; **capsules** immersed or shortly exerted, erect and symmetrical; peristome double but endostome greatly reduced, or single and exostomial. **Calyptrae** cucullate or mitrate, naked or hairy, roughened.

43. Phyllogonium viscosum (P. Beauv.) Mitt. (**Fig. 41**)

Plants stout, in very shiny, golden-brown or green pendent masses to 30 cm long. **Leaves** crowded and overlapping in two rows, deeply concave-cymbiform but only moderately complanate, usually irregularly undulate when dry, oblong-ovate, shortly apiculate, 3–5 mm long. **Setae** 2.5–3.5 mm long; **capsules** up to 2 mm long. **Spores** 21–25 µm diameter.

Note.—Reported as occurring in Madre de Dios near Shintuya (elev. 550 m) from a collection by *René Chavez 857* at MO. We have not seen the collection.

Habitat.—Typically montane forests to shrubby páramo, 300–3600 m (although quite rare below 1000 m), pendent from trunks and branches of trees.

PILOTRICHACEAE

Plants medium-sized to robust, sometimes small, in pale-green to golden or reddish mats or tufts. **Stems** mostly creeping, sometimes erect, scarcely branched to regularly bipinnate, often complanate foliate. **Lateral and dorsal/ventral leaves** often differentiated in shape, areolation, and costal development, sometimes isophyllous, sometimes contorted when dry, lanceolate to broadly ovate; margins not or sometimes bordered by narrow elongate cells, mostly serrulate to serrate, rarely entire, plane or less often erect or narrowly recurved; **costa** mostly strong and double, sometimes short and double, rarely absent; cells isodiametric to linear, smooth, prorulose or papillose, thin- to thick-walled; alar cells not differentiated. **Setae** usually elongate, sometimes short, smooth or roughened to spinose-papillose; **capsules** usually inclined, if erect then setae short and/or peristome reduced; annulus differentiated or not; opercula conical-rostrate; peristome double. **Calyptrae** mitrate, lobed, not or rarely fringed (*Crossomitrium*), often hairy, the hairs usually erect, often ± roughened. **Spores** spherical, mostly finely papillose.

1. Costa strong, ending near apex ______ **Callicostella pallida**
1. Costa weak or lacking, ending near or below midleaf.
    2. Costa lacking ______ **Crossomitrium patrisiae**
    2. Costa weak, from 1/4 to 1/2 leaf length.
        3. Plants not or obscurely complanate-foliate, mostly erect, usually in tufts on twigs and branches, densely tomentose below ______ **Lepidopilum surinamense**
        3. Plants mostly strongly complanate-foliate, usually mat-forming, on various substrates, not or scarcely tomentose at base of stems ______ **Lepidopilum affine**

CALLICOSTELLA (Müll. Hal.) Mitt.

Plants small to medium-sized, in pale green to golden, extensive, flat mats. **Stems** creeping, freely but irregularly branched, complanate-foliate. **Leaves** often contorted when dry, densely to laxly foliate, **lateral and dorsal leaves** differentiated from **ventral leaves**, leaves mostly oblong to oblong-ovate, gradually acuminate to more often broadly rounded to truncate and apiculate, ± asymmetrical; margins serrulate to coarsely serrate above, rarely subentire to crenulate, plane, not or obscurely bordered; **costa** double, often ending near leaf apex with apices connivent, often toothed, especially at the apex; cells short, mostly 1–2:1, unipapillose, weakly prorulose or smooth, thin- to thick-walled, porose or not. **Setae** elongate, short to very long, smooth throughout to roughened-papillose above or throughout, reddish, flexuose; **capsules** mostly inclined to pendent, rarely erect, ovoid to cylindrical, usually with a well-defined neck; exostome teeth reddish, cross striolate below, coarsely papillose above; endostome with a medium-high basal membrane, segments papillose and/or spiculose. **Calyptrae** naked or with a few hairs.

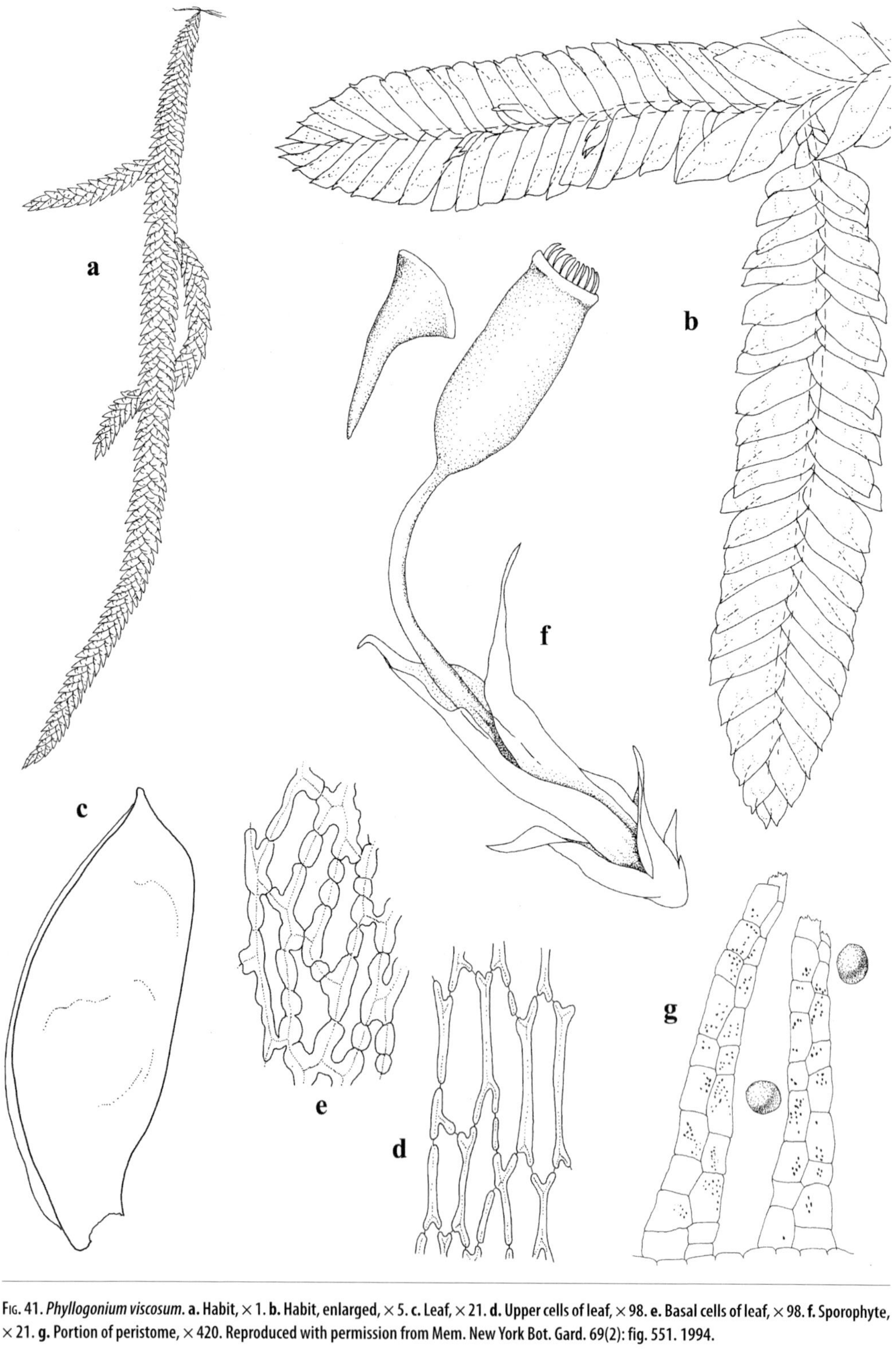

Fig. 41. *Phyllogonium viscosum*. **a.** Habit, × 1. **b.** Habit, enlarged, × 5. **c.** Leaf, × 21. **d.** Upper cells of leaf, × 98. **e.** Basal cells of leaf, × 98. **f.** Sporophyte, × 21. **g.** Portion of peristome, × 420. Reproduced with permission from Mem. New York Bot. Gard. 69(2): fig. 551. 1994.

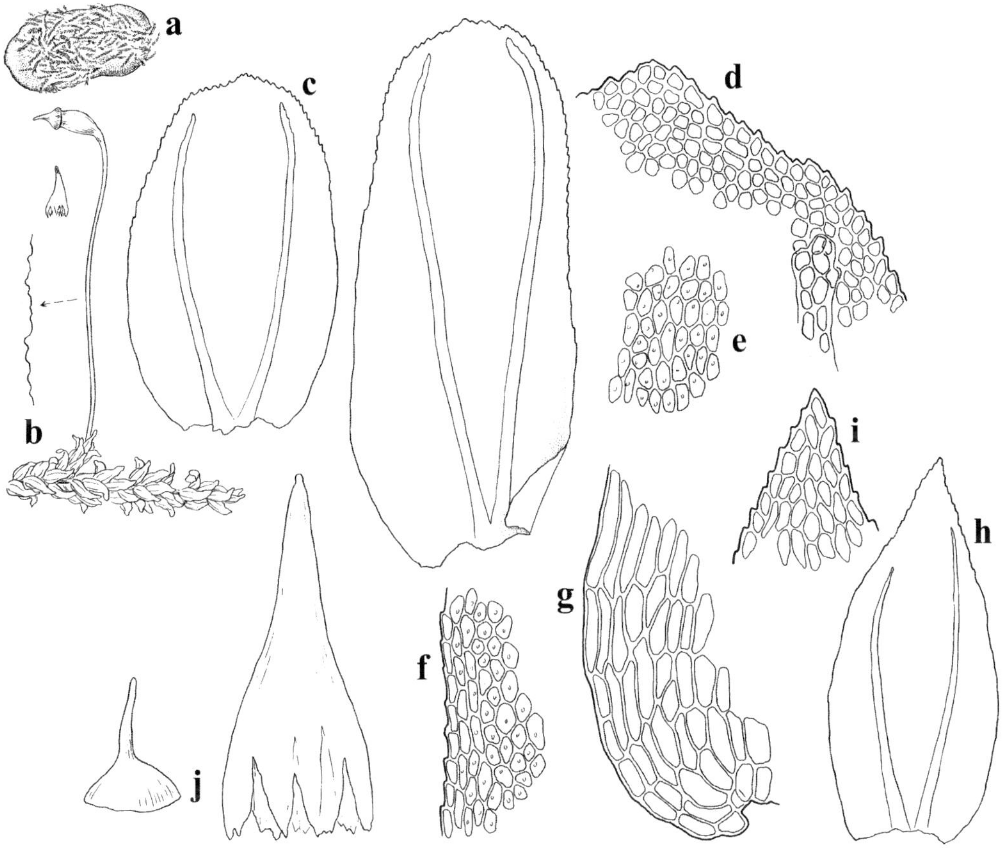

FIG. 42. *Callicostella pallida*. **a.** Habit. **b.** Stem apex and sporophyte with calyptra, showing magnification of seta profile. **c.** Lateral leaves. **d.** Lateral leaf apex, showing costa apex. **e.** Lateral leaf laminal cells, midleaf. **f.** Lateral leaf margin, midleaf. **g.** Basal marginal lateral leaf cells. **h.** Ventral leaf. **i.** Ventral leaf apex. **j.** Operculum and calyptra. Reproduced with permission from Mem. New York Bot. Gard. 82: plate 23, figs. 9–18. 1998.

44. **Callicostella pallida** (Hornsch.) Ångstr. (**Fig. 42**)

Plants medium-sized, in lustrous or dull, green, golden, or brownish, mostly dense mats. **Stems** to ca. 1.5–4 cm long, subpinnately to irregularly branched, densely foliate. **Leaves** contorted when dry; **lateral leaves** erect-spreading, **dorsal and ventral leaves** erect, oblong to oblong-ovate, typically broadly rounded at the apex, often apiculate, sometimes acute, rarely short-acuminate, 0.7–1.5 × 0.5–0.6 mm; margins typically serrulate to serrate above, with ± swollen and often bifid teeth, plane, not or obscurely bordered below; **costa** mostly subpercurrent, ending ca. 5 cells below apical margins, the upper 1/4 typically converging; cells ± isodiametric, ca. 9 µm wide, usually with a single papilla over each lumin, rarely smooth, thin- to firm-walled, at midleaf sometimes becoming hexagonal, ca. 2–3.1, otherwise like upper ones, becoming short-rectangular, smooth, and often colored across the insertion. **Setae** smooth throughout or roughened above to throughout, reddish, 0/6–1.5 cm long, flexuose, curved at extreme apex; **capsules** inclined to pendent, ovoid to cylindrical, 0.7–1.3 mm long, often with a roughened neck.

Habitat.—Very common in non-flooded moist lowland and lower montane forest, 150–1100 m, on soil, humus, wood, and rock.

CROSSOMITRIUM Müll. Hal.

Plants small to medium-sized, in lustrous, bright to pale green to golden brown, mostly flat, dense to lax mats. **Stems** creeping, little branched, the branches not or somewhat differentiated, ± complanate-foliate. **Leaves** sometimes ± contorted when dry, sometimes falcate, somewhat distant to crowded or overlapping; lateral and dorsal leaves differentiated in shape and orientation, wide-spreading to squarrose, oblong-ovate to oblong-lanceolate, ovate or ± orbicular, obtuse to acuminate, the apex often carinate; margins serrulate, often to base, the teeth usually at cell ends, often bifid, plane, not bordered; **costa** none; cells long-rhomboidal to linear, smooth, thin-walled, sometimes shorter at the extreme apex. **Setae** elongate, roughened above or throughout, reddish, often curved above; **capsules** erect to suberect, short-cylindrical, symmetrical; exostome teeth papillose throughout; endostome with a low basal membrane, segments slenderly keeled, narrowly perforate, about as long as the teeth. **Calyptrae** naked, lobed at base, smooth or ± roughened above.

45. Crossomitrium patrisiae (Brid.) Müll. Hal. (**Fig. 43**)

Plants medium-sized, in lustrous, loosely adnate, bright green to golden brown, often dense mats. **Stems** to 8 cm long, little-branched. **Leaves** ± contorted when dry, often falcate, lateral leaves wide-spreading to squarrose, oblong-lanceolate to broadly oblong-ovate, widest below midleaf, asymmetrical, 0.85–2.1 × 0.2–0.8 mm; margins serrulate almost to base, sometimes bifid, with occasional papillae above or not, plane; cells long-rhomboidal to linear, 50–160 × 8.5–14 µm. Asexual propagula common in pads on bases of leaves, and on rhizoids. **Setae** roughened above, smooth below, reddish, 6–12 mm long, curved; **capsules** suberect to pendent, short-cylindrical, 0.5–1.75 mm long.

Habitat.—Common in moist lowland to lower montane forest, 100–1000 m, on leaves of vascular plants and, to a lesser extent, on twigs, old wood, and humus.

LEPIDOPILUM (Brid.) Brid.

Plants mostly medium-sized, sometimes smaller or robust, in mostly lustrous, bright green or less often golden or reddish, dense or lax mats or tufts. **Stems** creeping, often short, often with eroded leaves, with erect primary branches, mostly complanate-foliate. **Leaves** sometimes contorted, dry or moist, laxly or densely foliate, lateral and dorsal leaves often differentiated in size, shape, and symmetry, mostly wide-spreading, oblong-lanceolate to broadly ovate or obovate, gradually to abruptly acute to acuminate, rarely obtuse; margins entire to serrate, plane or narrowly recurved, sometimes bordered by elongate, narrow cells; **costa** double, often slender, mostly ending at midleaf or below; cells long-hexagonal to fusiform or linear, the apical ones sometimes shorter than those at midleaf, smooth, thin- or firm-walled, becoming shorter and often colored in 1–2 rows across the insertion. Asexual propagula common in leaf axils. **Setae** short to elongate, typically papillose or spinose-papillose throughout, rarely smooth below, reddish; **capsules** erect, symmetrical, cylindrical; exostome teeth papillose throughout; endostome with a low basal membrane, segments erect, about as long as the teeth. **Calyptrae** hairy, smooth, or roughened at the apex.

46. Lepidopilum affine Müll. Hal. (**Fig. 44**)

Lepidopilum ambiguum Broth.
Lepidopilum subobtusulum Broth.

Plants medium-sized, in lustrous, pale to dark green, dense mats. **Stems** short, with spreading, mostly simple branches, to ca. 2.5 cm long, complanate-foliate. **Leaves** sometimes crispate to undulate when dry, ± flat when moist, crowded, ± oblong, obtuse to apiculate, 1.5–2.5 × 0.5–1 mm; margins serrate above, plane, not bordered; **costa** ending 1/3–1/2 the leaf length; cells long-fusiform to linear, ± thin-walled, 80–150 × 12–14 µm, in apex rhomboidal, 2–6:1, becoming long-rectangular across the insertion in 1–3 rows. **Setae** papillose throughout, 6–12 mm long; **capsules** 1–1.5 mm long. **Calyptrae** sparsely hairy.

Habitat.—In non-flooded lowland forests, 200–400 m, on tree branches.

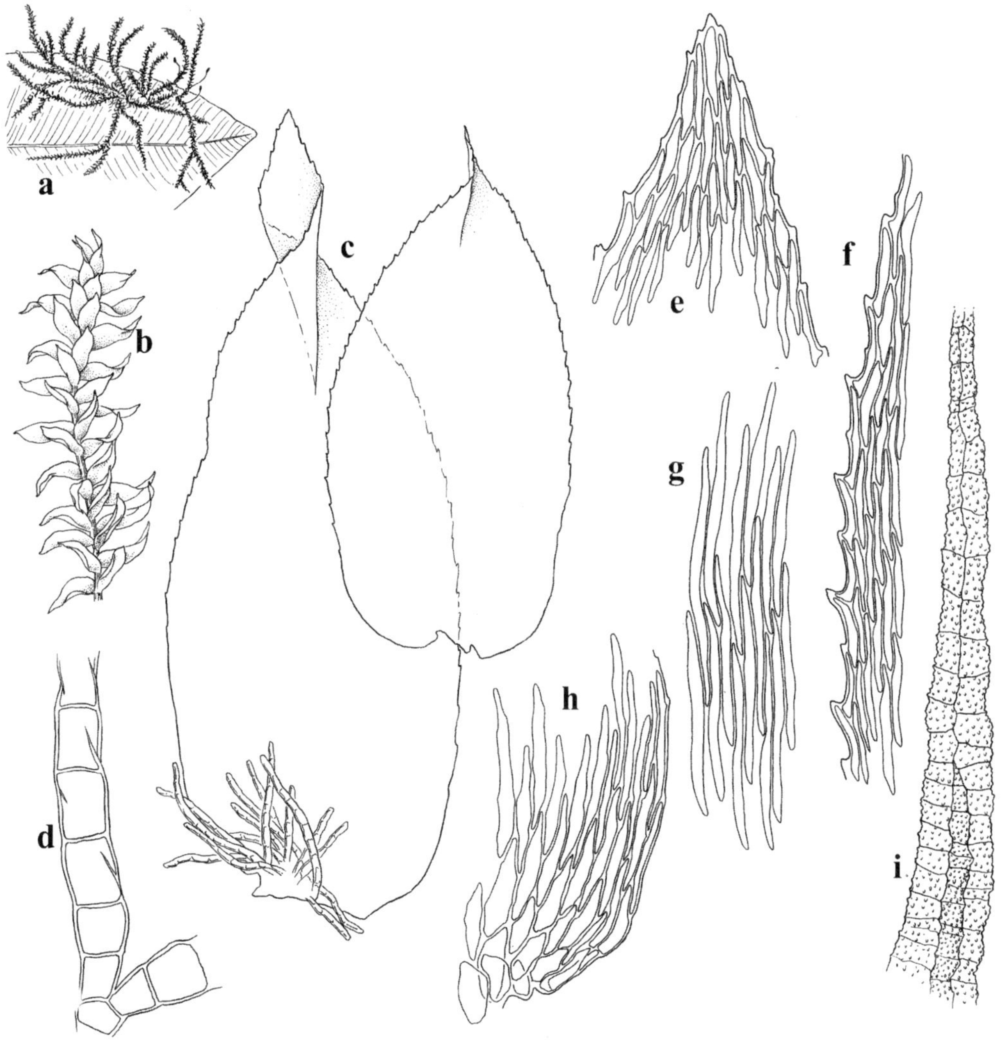

Fig. 43. *Crossomitrium patrisiae*. **a.** Habit. **b.** Stem apex. **c.** Leaves. **d.** Portion of propagulum. **e.** Leaf apex. **f.** Upper leaf margin. **g.** Laminal cells, midleaf. **h.** Basal marginal cells. **i.** Exostome tooth. Reproduced with permission from Mem. New York Bot. Gard. 82: plate 17, figs. 7–15. 1998.

47. Lepidopilum surinamense Müll. Hal. (Fig. 45)

Plants medium-sized, in lustrous, pale to dark or yellowish green, lax tufts. **Stems** short, densely tomentose, with erect, mostly simple branches, to ca. 6 cm long, weakly or not complanate-foliate. **Leaves** mostly contorted when dry, ± flat when moist, crowded below, distinct above, narrowly oblong, oblong-ovate or obovate, abruptly short-acuminate to apiculate, the apex often pinched and twisted, 2.5–3.2 × 1.5–1.8 mm; margins serrate to serrulate in upper 2/3–3/4, plane, weakly to distinctly bordered by 1–2 rows of elongate, narrow cells; **costa** 1/3–1/2 the leaf length; cells rhomboidal to hexagonal, 60–100 × 15–35 µm, often thick-walled and porose, becoming oblong to rectangular and colored across the insertion in 1–3 rows. Asexual propagula clustered in leaf axils, 10–15 celled. **Setae** smooth or roughened above, 2.5–3.0 mm long; **capsules** 2.5–3.0 mm long. **Calyptrae** sparsely hairy.

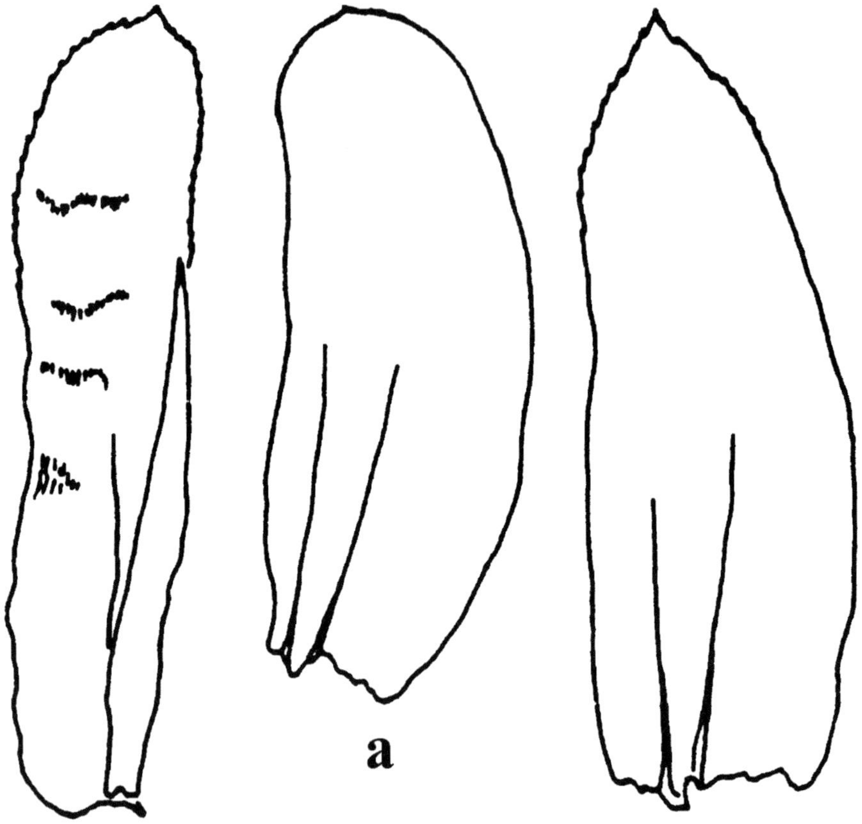

Fig. 44. *Lepidopilum affine.* **a.** Leaves, showing variation. Reproduced with permission from Mem. New York Bot. Gard. 76(3): fig. 92. 2003.

Habitat.—In non-flooded moist lowland forest and lower montane forest, 200–1500 m, typically on twigs, less often on tree trunks and logs.

POTTIACEAE

Plants small to robust, typically caespitose, dull, usually green above, brown, yellow-brown, or reddish below. **Stems** generally erect, radulose below, occasionally tomentose. **Leaves** usually larger and more crowded at the stem tip, often contorted when dry, spreading when moist, lanceolate or sometimes oval to ligulate, the base often differentiated; margins usually entire, plane to revolute or involute; **costa** single, subpercurrent to long excurrent, upper cells occasionally bistratose, small, subquadrate, usually papillose, occasionally mammillose-bulging or smooth; basal cells usually enlarged, often thin-walled. **Setae** usually elongate; **capsules** usually erect and symmetrical, spherical, ovoid or cylindrical; annulus often differentiated; opercula conical to rostrate; peristome generally present, single, consisting of 32 filiform divisions or 16 various perforate or cleft teeth. **Calyptrae** nearly always cucullate, smooth or rarely roughened, naked. **Spores** mostly spherical, finely papillose.

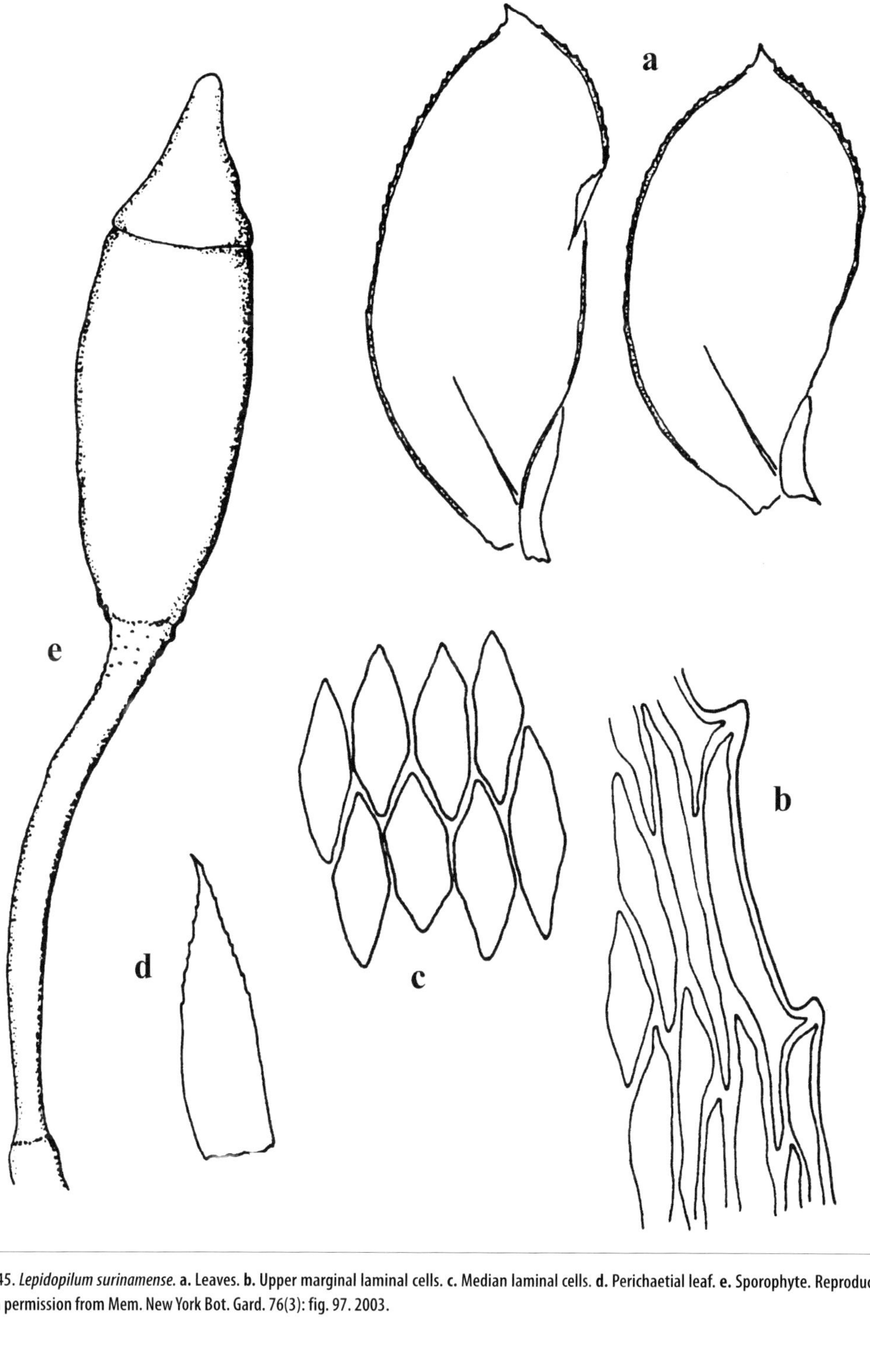

Fig. 45. *Lepidopilum surinamense*. **a.** Leaves. **b.** Upper marginal laminal cells. **c.** Median laminal cells. **d.** Perichaetial leaf. **e.** Sporophyte. Reproduced with permission from Mem. New York Bot. Gard. 76(3): fig. 97. 2003.

BARBULA Hedw.

Plants in loose or dense, green, yellow-brown or blackish tufts. **Stems** erect, sparsely and irregularly branched. **Leaves** oblong-spathulate to lanceolate or long-triangular, broadly concave or narrowly channeled, rounded or acute at the apex, weakly differentiated to sheathing at the base; margins often recurved, entire to slightly dentate; **costa** subpercurrent to short-excurrent; upper cells subquadrate to short-rectangular, generally papillose; basal cells usually differentiated, rectangular, and smooth or somewhat papillose. **Setae** elongate; **capsules** short- to long-cylindrical; annulus usually present, occasionally revoluble; opercula short- to long-conical; peristome weakly to strongly twisted, consisting of 32 filiform, spiculose teeth. **Calyptrae** cucullate.

48. Barbula indica (Hook.) Spreng. ex Steud. (**Fig. 46**)

Plants yellow-green to brown. **Stems** simple to few branched, up to 25 mm tall; propagula often present, axillary, stalked, green to brown, obovoid, clavate, spindle-shaped, ellipsoidal or globose, consisting of 3–50 cells. **Leaves** crowded to ± distant, nearly uniform in size, when dry erect-incurved, often infolded above, when moist erect-spreading, straight or incurved, 0.5–2 mm long, oblong-ovate to ligulate, occasionally lanceolate or long triangular, grooved at the costa, rounded to bluntly acute, apiculate by a single pellucid cell, the base scarcely differentiated to much broadened and square, rectangular or oval; margins plane to slightly recurved at midleaf, entire; **costa** subpercurrent to percurrent as a apiculate apex or occasionally excurrent as a sharp mucro; cells above quadrate, 7–10 µm wide, with 4–8 low, mostly hollow papillae per cell; basal cells 8–12 µm wide, 2–5:1, rectangular. **Setae** 7–13 mm long, red-brown; **capsules** red-brown, ellipsoidal to ovoid, erect, occasionally curved; opercula short- to long-conical; peristome divisions ± fused in pairs, with many articulations, red to orange. **Spores** smooth to weakly papillose, 8–12 µm diameter.

Notes.—This species typically has ligulate leaves that are deeply grooved along the costa on the upper ventral surface, weakly recurved leaf margins, rounded to bluntly acute and apiculate leaf apices, round, densely papillose cells on the upper, dorsal surface of the costa, and when present numerous axillary propagula.

Habitat.—Lowland to humid or semi-dry lower montane forests, 0–3900 m, on soil, limestone, rock, or walls, on banks of roads or streams, sometimes on bark.

DOLOTORTULA R.H. Zander

Plants small, gregarious or in mats. **Stems** erect, unbranched; rhizoids sparsely scattered on stem. **Leaves** oblong, oblong-ligulate or oblong-spathulate, twisted and contorted when dry, spreading when wet, rounded to emarginate, bluntly apiculate; lamina unistratose; margins with a narrow, red, multistratose border of cells ending just before the apex; **costa** slender, subpercurrent to percurrent; leaf cells smooth and thin-walled, upper cells hexagonal to short-rectangular, basal cells longer rectangular. **Setae** long, red-brown; **capsules** cylindrical, erect, smooth; annulus present; opercula erect, long-rostrate; peristome teeth 32, erect to weakly twisted. **Calyptrae** cucullate. **Spores** spherical, lightly papillose.

49. Dolotortula mniifolia (Sull.) R.H. Zander (**Fig. 47**)

Plants in loose or dense, brownish-green tufts 2–3 mm tall in depauperate forms or to 8 mm in mesic or hydric forms. **Leaves** crisped and twisted when dry, loosely spreading when moist, 1–4 × 0.75–2 mm, oblong, oblong-lingulate or obovate, narrowed at the insertion, emarginate, rounded or acute; margins plane, entire or slightly denticulate at the apex; **costa** slender, brown or yellow, smooth, ending a few cells below the apex or percurrent; upper cells 20–40 × 15–30 µm, irregularly rectangular to hexagonal, thin-walled, smooth; marginal cells yellow to red, linear, thick-walled, 4–6 stratose; basal cells gradually differentiated. **Setae** 8–12 mm long, red; **capsules** ca. 2 mm long, straight, with an abrupt neck, red; peristome red, straight or slightly twisted. **Spores** 11–13 µm diameter.

Habitat.—Lowland to lower montane forests, 0–2200m, on soil along streams, disturbed banks, or exposed rocks.

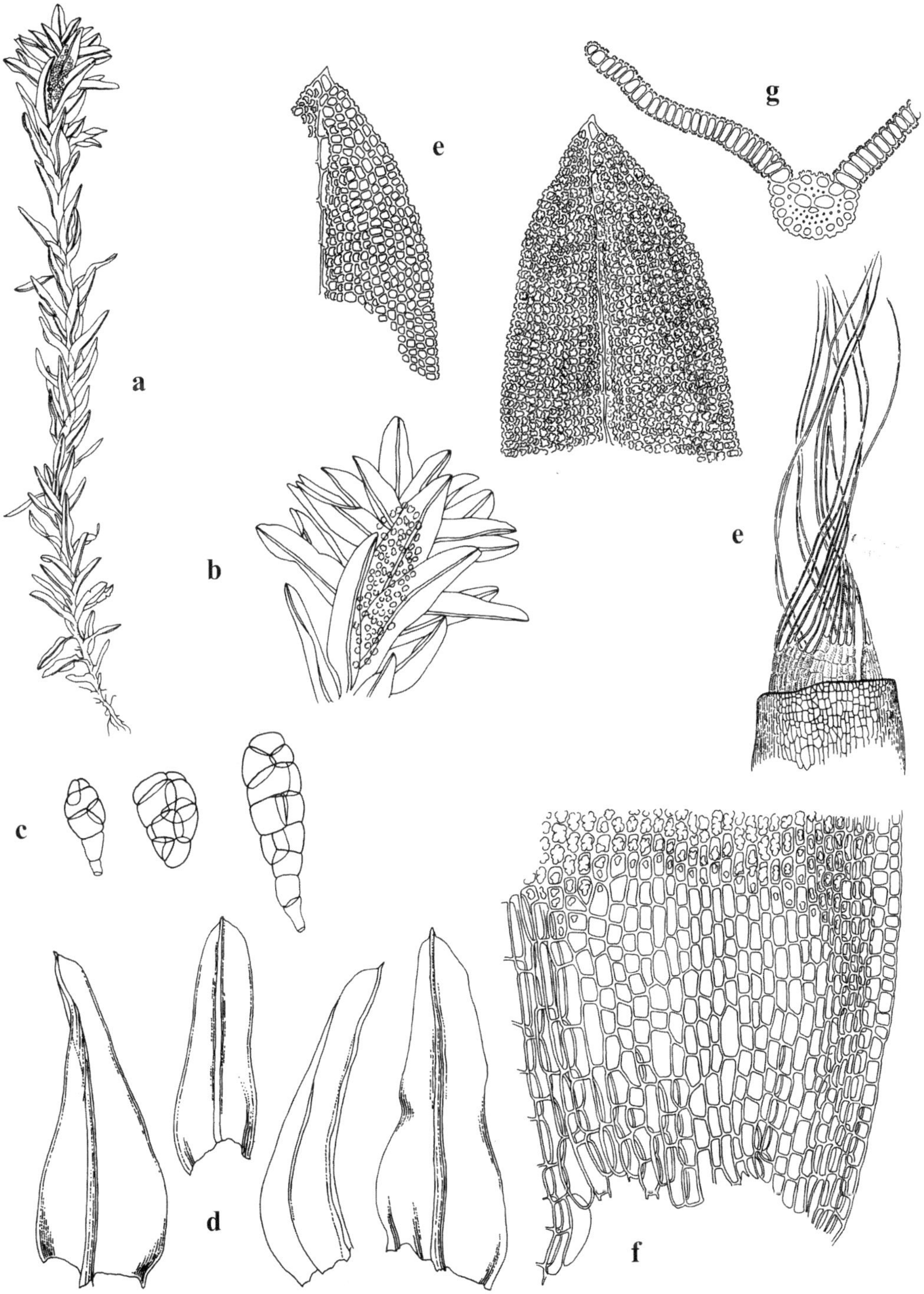

FIG. 46. *Barbula indica* var. *indica*. **a.** Habit, × 8. **b.** Portion of plant showing masses of gemmae, × 26. **c.** Gemmae, × 309. **d.** Leaves, × 59. **e.** Apical cells of leaf (with and without indication of papillae), × 309. **f.** Basal cells of leaf, × 309. **g.** Leaf section, × 309. Reproduced with permission from Mem. New York Bot. Gard. 69(1): fig. 215. 1994.

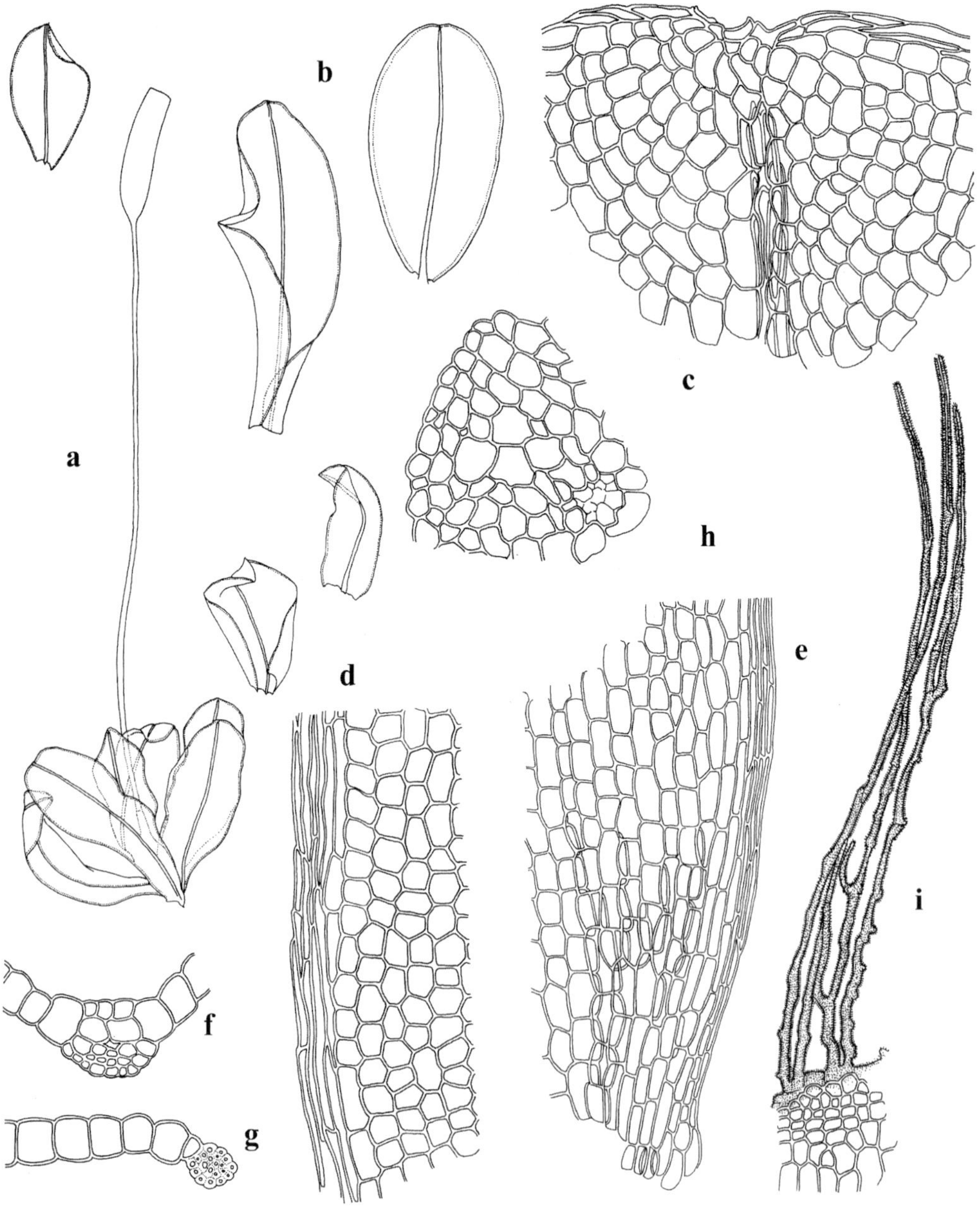

FIG. 47. *Dolotortula mniifolia*. **a.** Habit, × 10. **b.** Leaves, × 20. **c.** Apical cells of leaf, × 285. **d.** Marginal cells of leaf, × 285. **e.** Basal cells of leaf, × 120. **f.** Costa in section, × 185. **g.** Leaf margin in section, × 285. **h.** Portion of stem in section, × 120. **i.** Peristome teeth, × 185. Reproduced with permission from Mem. New York Bot. Gard. 69(1): fig. 236. 1994.

PTEROBRYACEAE

Plants medium to robust, often regularly pinnate, generally frondose above a stipitate base. **Primary stems** creeping, filiform, with small, clasping, scalelike, fragile leaves; **secondary stems** erect or pendent, simple or irregularly branched to regularly 1- or 2-pinnate. Gemmae common in leaf axils, cylindrical, septate. **Leaves** erect-spreading, spreading or squarrose, smooth or plicate, concave, oblong to ovate or ovate-lanceolate, occasionally auriculate or decurrent, acute to acuminate; margins entire to serrulate; **costa** single, short and double, or absent; upper cells rhomboidal to linear (or rarely rhombic to short-elliptic), thick-walled, often porose, smooth or papillose because of projecting upper ends; basal cells colored, strongly thick-walled, and porose; alar cells often distinctly differentiated. **Setae** ± short (rarely elongate), smooth throughout or rough above; **capsules** immersed to exserted, erect, oblong to ovoid; opercula conical-rostrate, peristome single or double, exostome teeth 16; endostome often rudimentary. **Calyptrae** cucullate or mitrate, often sparsely pilose. **Spores** large, papillose.

1. Costa subpercurrent ______ **Pireella pohlii**
1. Costa ending well below apex ______ **Henicodium geniculatum**

HENICODIUM (Müll. Hal.) Kindb.

50. Henicodium geniculatum (Mitt.) W.R. Buck (**Figs. 48, 66**)

Plants medium-sized, in loose, dull or somewhat shiny, yellow-green or brownish mats. **Secondary stems** erect-ascending, straight or curved, simple or sparsely branched, up to 3 cm long, often bearing filiform, septate brood bodies in leaf axils. **Leaves** crowded, loosely appressed and wrinkled or ± plicate when dry, erect-spreading when moist, somewhat concave, 1–2 mm long, oblong-ovate or oblong-lanceolate, acute or gradually acuminate; margins revolute nearly to the apex, entire or crenulate at the apex; **costa** single, slender, ending at or above midleaf; cells linear-flexuose, with 1 or sometimes 2–4 papillae on both surfaces or rarely smooth, small and subquadrate in numerous rows at the basal angles. **Setae** 4–6 mm long, straight or slightly flexuose, smooth; **capsules** exerted, erect and symmetrical, oblong-cylindrical ca. 2 mm long; exostome with a well developed prostome; endostome smooth, with a low basal membrane, perforate, almost as long as the teeth.

Notes.—This species is distinguished by mostly unbranched, erect secondary stems with revolute-margined leaves. The leaf cells are generally unipapillose and the costa ends well below the leaf apex. Gemmae are regularly produced on older secondary stems in large clusters in upper leaf axils.

Habitat.—Lowland to lower montane often open, mesic to dry forest, 200–1000 m, usually on twigs or less often on tree trunks.

PIREELLA Cardot

Plants small to robust, scattered or loosely tufted, green or yellow-green. Gemmae frequent. **Secondary stems** irregularly branched to pinnately frondose above a stipitate base. **Stipe leaves** appressed to squarrose, lanceolate-acuminate; **secondary stem and branch leaves** erect-spreading, concave, ovate to oblong-lanceolate, acute to acuminate; **costa** single; cells rounded-oval to rhombic or oblong-linear, thick-walled, smooth or papillose at back because of projecting ends; alar cells poorly differentiated or quadrate in few to several rows. **Setae** short or elongate, smooth or ± papillose above; **capsules** often immersed but occasionally exerted, oblong-ovoid; peristome double. **Calyptrae** cucullate or mitrate, mostly pilose, smooth. **Spores** spherical.

51. Pireella pohlii (Schwägr.) Brid. (**Fig. 49**)

Plants medium-sized, in green to golden, thin to dense, epiphytic colonies. Asexual propagula of 3–12-celled gemmae clustered in leaf axils and of microphyllous flagellate branches. Stems erect, frondose, usually stipitate, irregularly pinnately branched, erect, sometimes with flagellate branches. **Stipe leaves** erect-appressed, oblong-ovate, ± abruptly short-acuminate, ca. 2 mm long. **Stem and branch leaves** erect-spreading to spreading, broadly lanceolate, gradually acuminate, strongly concave, not plicate, 1–2 mm long; margins finely serrulate throughout, plane; **costa** single, subpercurrent; cells short-linear, often prorulose at upper ends, becoming broader, thicker-walled and porose towards the insertion; alar cells few, subquadrate, confined to the base of the auricle. **Setae** ca. 10 mm long; **capsules** exerted, ca. 2.5 mm long; exostome teeth smooth, often perforate at base; endostome rudimentary. **Calyptrae** cucullate, pilose.

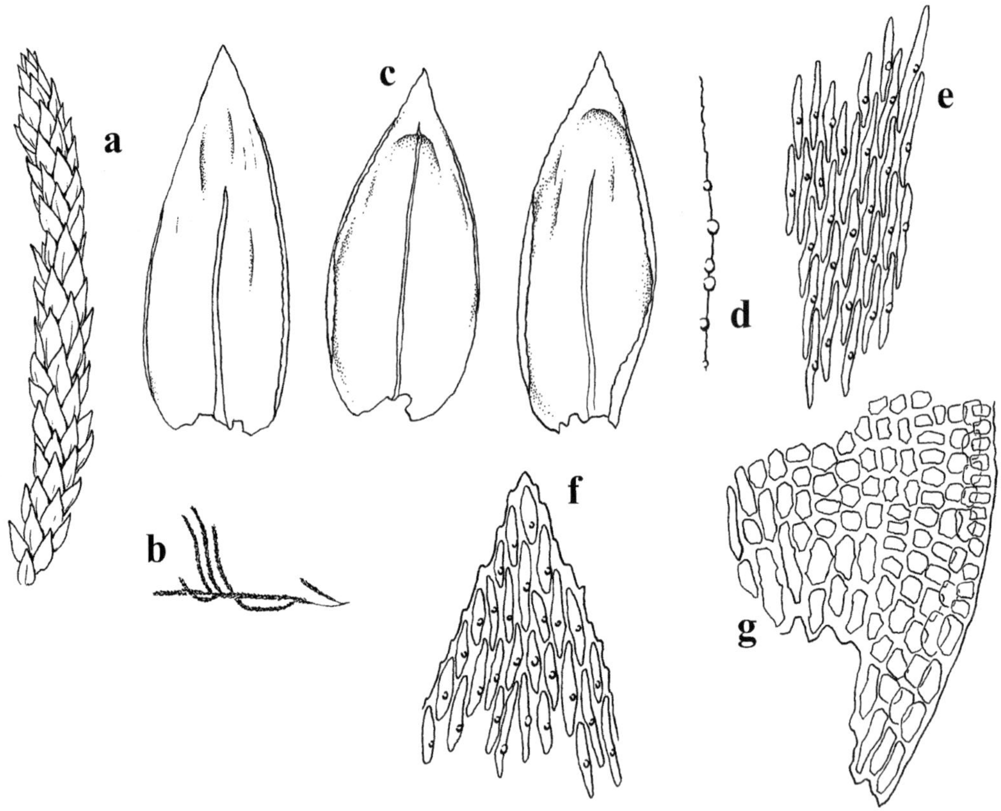

FIG. 48. *Henicodium geniculatum*. **a.** Portion of secondary stem. **b.** Habit. **c.** Secondary stem leaves. **d.** Leaf in profile, showing papillae. **e.** Laminal cells, midleaf. **f.** Secondary stem leaf apex. **g.** Basal marginal secondary stem leaf cells. Reproduced with permission from Mem. New York Bot. Gard. 82: plate 60, figs. 1–7. 1998.

Habitat.—In humid lowland to lower montane forest, 100–1000 m, on tree trunks, branches, and limestone.

SEMATOPHYLLACEAE

Plants small to robust, in often lustrous, soft to stiff, green to yellow-brown but typically golden-green, sometimes reddish, lax or dense mats. **Stems** creeping or rarely erect, irregularly branched or rarely ± regularly pinnate, sometimes complanate-foliate. **Leaves** of stems and branches mostly similar, occasionally differentiated, often subcomplanate, sometimes homomallous to falcate-secund, narrowly lanceolate to ovate to suborbicular, symmetrical, obtuse to acuminate, often concave, not or scarcely decurrent; margins not or rarely bordered, entire to serrate, plane, recurved or reflexed; **costa** short and double or absent; cells rhomboidal to linear-flexuose, typically long-hexagonal, smooth or papillose, seldom prorulose, mostly firm-walled; alar cells mostly enlarged, inflated or not, colored, occasionally quadrate and not inflated. **Setae** elongate, smooth or roughened above or rarely throughout, usually reddish; **capsules** erect to pendent, mostly asymmetrical, arcuate or straight, ovoid to cylindrical, smooth or rarely with a roughened neck, often constricted below the mouth when dry and empty; annulus usually not differentiated; opercula typically obliquely long-rostrate, rarely conical; peristome typically double, the endostome rarely lacking, exostome teeth typically strongly shouldered and bordered, on the front surface with a zig-zag median line or a median furrow, cross-striolate below, coarsely papillose above, sometimes reduced and papillose or

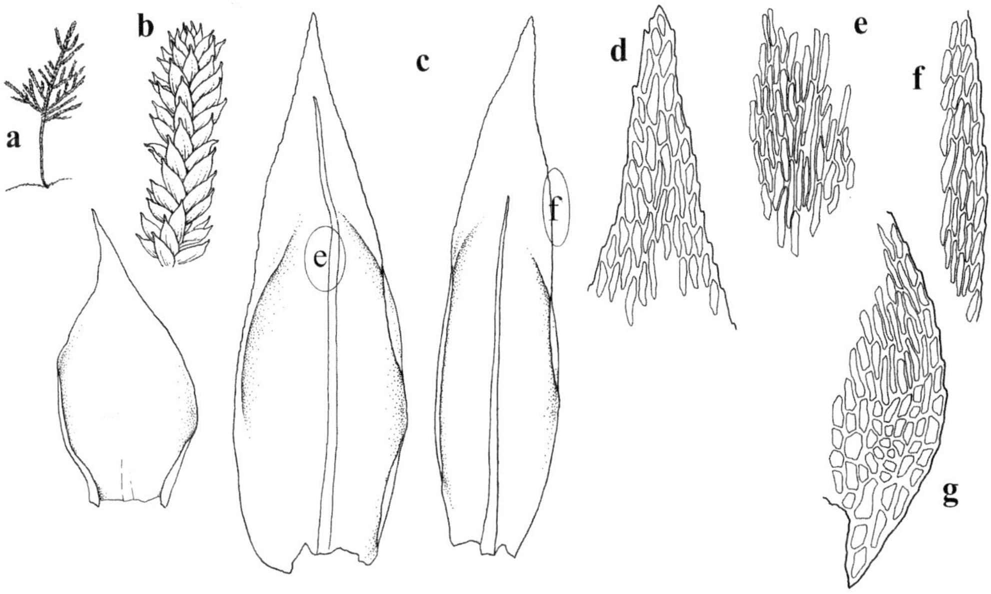

FIG. 49. *Pireella pohlii.* **a.** Habit. **b.** Branch. **c.** Stipe leaf (far left) and branch leaves. **d.** Branch leaf apex. **e.** Juxtacostal laminar cells. **f.** Branch leaf margin. **g.** Basal marginal branch leaf cells. Reproduced with permission from Mem. New York Bot. Gard. 82: plate 55, figs. 1–7. 1998.

smooth throughout. **Calyptrae** cucullate or rarely mitrate, naked or rarely sparsely hairy, usually smooth, sometimes scabrous above. **Spores** spherical, often papillose.

1. Median leaf cells papillose.
  2. Leaf cells pluripapillose, papillae several in a row ________ **Taxithelium planum**
  2. Leaf cells with a single papillae over cell lumen ________ **Trichosteleum fluviale**
1. Median leaf cells smooth.
  3. Leaves ligulate, obtuse-rounded ________ **Potamium lonchophyllum**
  3. Leaves broadly to narrowly lanceolate or ovate, acute to obtuse-rounded or short to long acuminate.
    4. Apical laminal cells short, 2–3:1 ________ **Sematophyllum subpinnatum**
    4. Apical laminal cells mostly longer, 6–12:1.
      5. Plants golden-green to brown, irregularly branched; alar cells mostly 2–4(–7) times larger than the quadrate cells directly above them ________ **Sematophyllum adnatum**
      5. Plants whitish green, ± regularly pinnate; alar cells mostly 5–10 times larger than the quadrate cells directly above them ________ **Sematophyllum subsimplex**

POTAMIUM Mitt.

52. Potamium lonchophyllum (Mont.) Mitt. (**Fig. 50**)

Plants medium-sized, dark-green to blackish green, forming loose mats. **Stems** and branches long and slender, radiculose at base of attachment. **Leaves** rather flaccid, distant, erect-spreading, narrowly oblong to ligulate, 1.6–2 mm long, bluntly acute; margins plane or folded on one side, reflexed at base, denticulate from about midleaf, distally serrulate; **costa** none; laminal cells smooth, median cells linear, apical cells shorter, rhomboidal to fusiform; basal cells oblong-rectangular, weakly porose, yellowish to golden; alar cells inflated, oblong or rectangular-rounded, 2–4 rows. **Setae** elongate; **capsules** pendent, urceolate-ovoid; peristome double, exostome teeth distinctly furrowed.

Habitat.—Wet to semi-dry tropical lowland forest, 120–240 m, submerged on rocks or tree roots in streams and on rocky or sandy soil along streams.

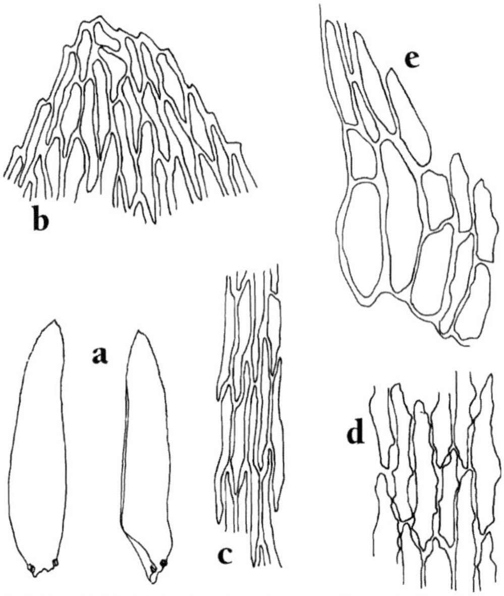

Fig. 50. *Potamium lonchophyllum*. a. Leaves. b. Upper cells. c. Upper marginal cells. Reproduced with permission from authors Churchill & Linares, 1995.

SEMATOPHYLLUM Mitt.

Plants small to robust, in mostly lustrous, ± yellow-green, often dense, extensive mats. **Stems** creeping, sparsely to freely but irregularly branched, the branches prostrate to erect, densely foliate, often curved. **Stem and branch leaves** similar, erect-spreading to homomallous to falcate-secund, lanceolate to ovate, acute to acuminate, rarely obtuse, usually concave; margins entire to serrulate above, subentire below, often reflexed; **costa** short and double or absent; cells rhomboidal to linear, sometimes flexuose, smooth or rarely unipapillose, often thick-walled, mostly not or only weakly porose, often becoming shorter in the apex if acute; alar cells enlarged and inflated, often in 2 rows in basal angles, usually yellow, the lower ones mostly oblong, the upper ones usually quadrate. **Setae** elongate, smooth, usually reddish; **capsules** mostly pendent, rarely erect to suberect, small, usually ± asymmetrical, ovoid to short-cylindrical; opercula mostly obliquely and slenderly long-rostrate; peristome double, exostome teeth strongly shouldered, strongly bordered, on the front surface with a zig-zag median line, cross-striolate below, coarsely papillose above. **Calyptrae** cucullate, naked, usually smooth. **Spores** spherical, finely papillose.

53. Sematophyllum adnatum (Michx.) E. Britton (**Fig. 51**)

Plants smallish, in pale- to golden-green, often flat, small, lax mats. **Stems** creeping, to ca. 3 cm long, freely but irregularly branched, the branches often ascending, short, somewhat curved. **Stem and branch leaves** similar, erect to erect-spreading, sometimes homomallous, lanceolate to oblong-lanceolate, 0.8–1.6 mm long, gradually acuminate, flat to shallowly concave; margins entire, often reflexed; cells long-rhomboidal to linear-flexuose, 6–12:1, smooth, firm-walled, not porose, not shorter in the acumen; alar cells enlarged, ± inflated, oblong, usually not more than 2–3 times as large as the quadrate cells above them, concolorous or colored across the insertion. **Setae** relatively short, smooth, reddish, 4–9 mm long, straight; **capsules** suberect, ± symmetrical, short-cylindrical, ca. 1 mm long, slightly constricted below the mouth when dry; annulus not differentiated or with a single row of cells fragmenting and usually falling with the opercula. **Spores** finely papillose, 10–17 µm diameter.

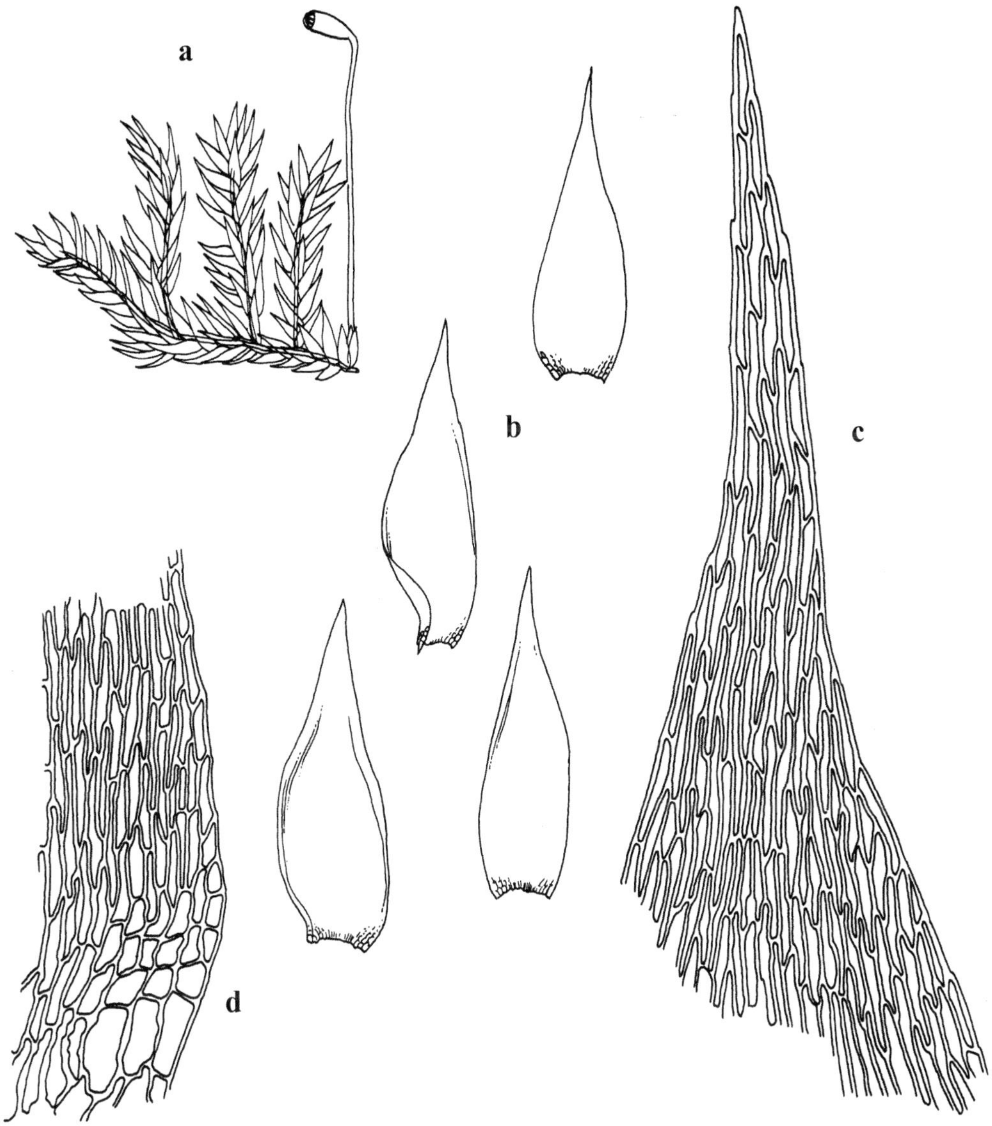

Fig. 51. *Sematophyllum adnatum.* **a.** Habit, × 10. **b.** Leaves, × 49. **c.** Apical cells of leaf, × 370. **d.** Basal marginal cells of leaf, × 370. Reproduced with permission from Mem. New York Bot. Gard. 69(2): fig. 744. 1994.

Habitat.- Wet lowland to lower montane forest, 0–1000 m, on tree trunks and logs, rarely on soil or rocks, in mostly mesic, open habitats.

54. Sematophyllum subpinnatum (Brid.) E. Britton (**Fig. 52**)

Plants small to medium-sized, in ± dull, golden-green, often dense mats. **Stems** creeping, to ca. 3 cm long, freely but irregularly branched, the branches often ascending, short, usually less than 1 cm long, blunt, curved. **Stem and branch leaves** similar, mostly homomallous, ovate to oblong-ovate, 0.75–1.2 m long, acute to abruptly short-acuminate, ± concave; margins entire, reflexed above, plane below; **costa** short and double or commonly absent; cells long-rhomboidal, 6–12:1, smooth, firm- to thick-walled, mostly not or

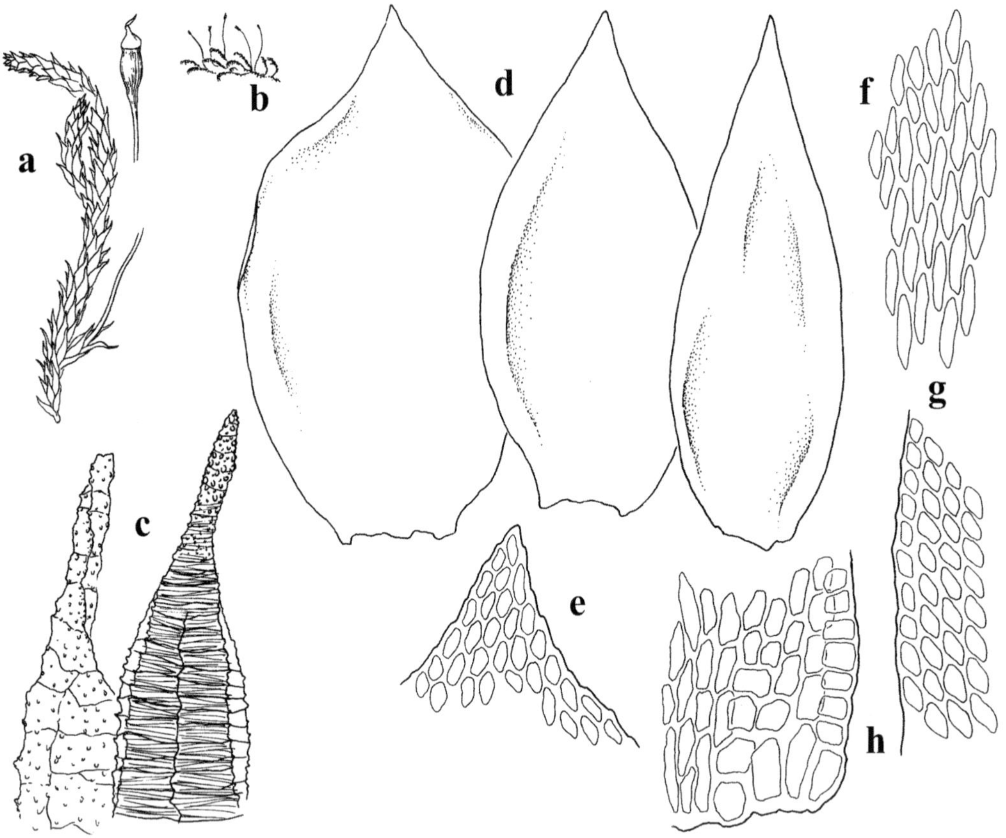

FIG. 52. *Sematophyllum subpinnatum*. **a.** Portion of stem and sporophyte. **b.** Habit. **c.** Portion of peristome. **d.** Leaves. **e.** Leaf apex. **f.** Laminal cells, midleaf. **g.** Leaf margin. **h.** Basal marginal leaf cells. Reproduced with permission from Mem. New York Bot. Gard. 82: plate 147, figs. 1–8. 1998.

only weakly porose, becoming shorter in the apex, rhomboidal to rounded-rhomboidal, 2–4:1; alar cells enlarged but not conspicuously inflated, colored across the insertion, oval to oblong, usually not more than 2–3 times larger than the quadrate cells above them. **Setae** elongate, orange to reddish brown, 0.5–1 cm long, straight or curved at base when dry; **capsules** erect to suberect, ± symmetrical, short-cylindrical, 1–1.5 mm long, often strongly constricted below the mouth when dry; annulus not differentiated. **Spores** finely papillose, 15–23 μm diameter.

Habitat.—Mesic to humid forest, 0–1500 m, growing mostly on tree trunks and logs, rarely on soil or rock.

55. Sematophyllum subsimplex (Hedw.) Mitt. (Fig. 53)

Plants slender, in ± lustrous, silky, pale-green to grayish green, often extensive, thin mats. **Stems** creeping, dark red, ca. 3–4(8) cm long, freely and often ± regularly pinnately branched, the branches wide-spreading, short, somewhat complanate-foliate. **Stem and branch leaves** similar, erect- to wide-spreading, not at all homomallous, lanceolate-ovate to ovate, stem leaves 0.65–1.3 mm long, branch leaves 0.55–1 mm long, gradually acuminate, occasionally the apices twisted in stem leaves, somewhat concave; margins entire or serrulate above, entire below, plane or recurved; cells linear to linear-flexuose, 10–20:1, usually smooth, rarely unipapillose, thick-walled, often porose, not shorter in the acumen; alar cells enlarged, inflated, *usually* deeply colored across the insertion, oval to oblong, 5–10 times as large as the subquadrate cells above them.

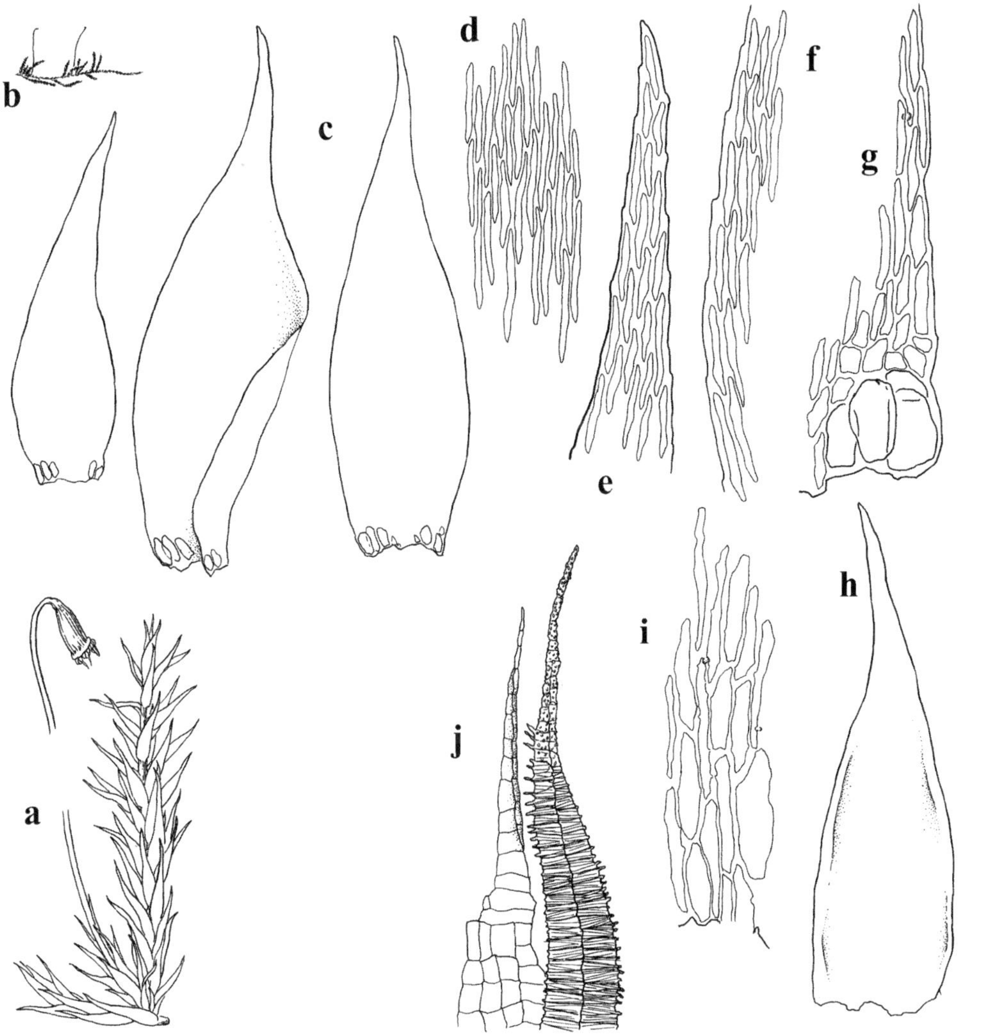

FIG. 53. *Sematophyllum subsimplex*. **a.** Portion of stem and sporophyte. **b.** Habit. **c.** Leaves. **d.** Laminal cells. **e.** Leaf apex. **f.** Leaf margin. **g.** Basal marginal leaf cells. **h.** Perichaetial leaf. **i.** Portion of peristome. Reproduced with permission from Mem. New York Bot. Gard. 82: plate 149, figs. 1–10. 1998.

Setae elongate, reddish, 1–1.8 cm long; **capsules** horizontal to pendent, asymmetrical, ovoid, 0.8–1.3 mm long; annulus not differentiated. **Spores** finely papillose, 14–24 µm diameter.

Habitat.—Lowland moist forests, from near sea level to 1200 m, on logs and less often on limestone or the bases of trees.

TAXITHELIUM Mitt.

Plants small to medium-sized, in somewhat dull to lustrous, yellow-green, thin, often extensive mats. **Stems** creeping, subpinnately branched, the branches usually complanate-foliate. **Stem and branch leaves** ± similar, the stem leaves more acuminate, somewhat larger and less papillose than branch leaves, all leaves erect- to wide-spreading, not falcate-secund; margins serrulate, usually plane; **costa** short and double or absent; cells linear, sometimes flexuose, mostly seriately pluripapillose, rarely prorulose or smooth, thin- to firm-walled, not or scarcely porose; alar cells few and quadrate in basal angles, or inflated and colored. **Setae** elongate, usually reddish; **capsules** inclined, asymmetrical, ovoid to cylindrical, constricted below the mouth when dry and empty; annulus not differentiated; opercula conical to obliquely conical-rostrate; peristome double, exostome shouldered, bordered, on the front surface with a zig-zag median line, cross-striolate below, coarsely papillose above. **Calyptrae** cucullate, naked, smooth. **Spores** spherical, finely papillose.

56. Taxithelium planum (Brid.) Mitt. (**Fig. 54**)

Plants slender to medium-sized, yellow-green, thin, often extensive, flat mats. **Stems** long-creeping, to ca. 15 cm long, but usually less than 10 cm, subpinnately branched, the branches prostrate to slightly ascending, complanate to terete-foliate. **Stem and branch leaves** ± similar, erect-spreading, crowded, complanate to terete, narrowly to broadly ovate, 0.6–1.2 mm long, obtuse or acute to short-acuminate, concave; margins serrulate nearly to base, sometimes doubly toothed, plane; **costa** short and double or absent; cells linear, seriately pluripapillose with 5–9 papillae over each lumen at back, the papillae small and blunt, sometimes variously branched, becoming shorter and smooth in the apex, becoming shorter, smooth, thick-walled, and porose toward the insertion; alar cells 2–4 large, inflated, occasionally colored, with small marginal group of quadrate, thin-walled cells above them. **Setae** elongate, reddish, 0.8–2 cm long, usually curved at base of urn; **capsules** inclined, asymmetrical, ovoid to short-cylindrical, ca. 1 mm long. **Spores** almost smooth or finely papillose, 10–15 µm diameter.

Habitat.—Wet lowland to lower montane forest, from near sea level to 1000 m, on most substrates, but most common on soil, logs, and base of trees.

TRICHOSTELEUM Mitt.

Plants small to moderately robust, in mostly yellow-green to golden, thin, often extensive mats. **Stems** creeping, freely but irregularly branched, the branches sometimes obscurely complanate-foliate. **Stem and branch leaves** similar, erect-spreading or falcate-secund, lanceolate to ovate-lanceolate, acute to acuminate-subulate, concave; margins subentire to serrate above, plane or sometimes recurved; **costa** short and double or absent; cells linear, unipapillose, usually thick-walled and porose, sometimes becoming shorter in the extreme apex, becoming thicker-walled and more conspicuously porose toward the insertion; alar cells greatly enlarged and inflated, oblong, often colored. **Setae** elongate but relatively short, reddish, curved just below the urn; **capsules** suberect to pendent, small, ± asymmetrical, ovoid to cylindrical; annulus not differentiated; opercula obliquely and slenderly long-rostrate; peristome double, exostome teeth shouldered, bordered, on the front surface with a zig-zag median line or a median furrow, cross-striolate below, coarsely papillose above. **Calyptrae** cucullate, naked or very rarely sparsely hairy, smooth, roughened, or ridged above. **Spores** spherical, papillose.

57. Trichosteleum fluviale (Mitt.) A. Jaeger (**Fig. 55**)

Plants medium-sized, in golden-green to yellow-brown, lax, often extensive mats. **Stems** to ca. 3 cm long, reddish, the branches usually short, obscurely complanate-foliate. **Stem and branch leaves** erect-spreading, broadly ovate-lanceolate, 0.75–1 mm long, acute, deeply concave; margins subentire to serrulate throughout, plane or occasionally recurved; cells linear, unipapillose in upper 2/3 of leaf, thin-walled, becoming rhombic toward the apex, becoming thicker-walled and porose toward the insertion, the papillae over most but not all cells; alar cells lightly colored, 2–3 at each basal angle. **Setae** elongate ca. 0.8–1 mm long; **capsules** ovoid-cylindrical, 0.7–0.9 mm long. **Spores** 12–15 µm diameter.

Habitat.—Humid lowland to lower montane forest, from near sea level to 600 m, on rocks, trunks of trees, or logs.

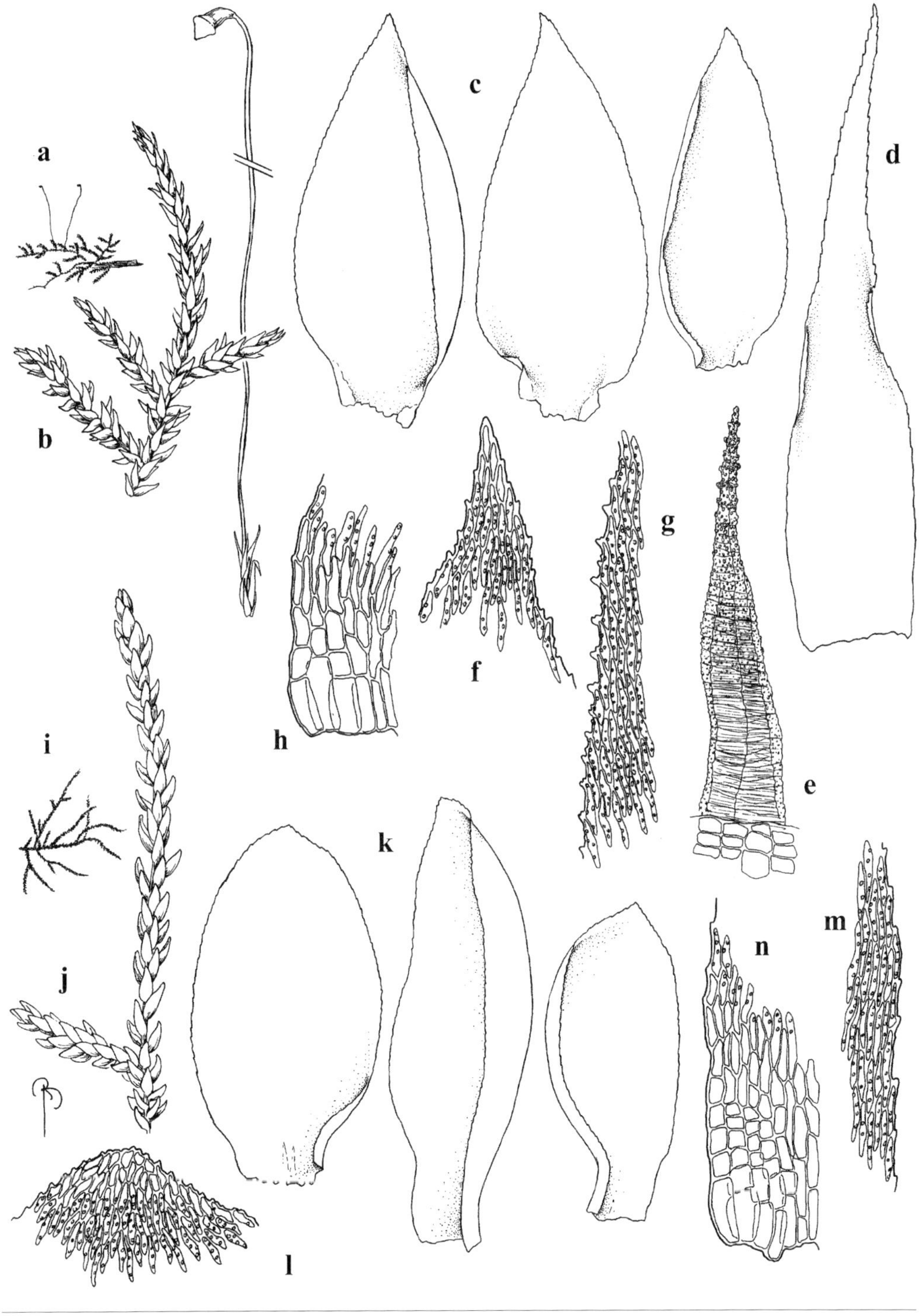

FIG. 54. *Taxithelium planum*. **a.** Habit. **b.** Portion of stem and sporophyte. **c.** Leaves. **d.** Perichaetial leaf. **e.** Exostome tooth. **f.** Leaf apex. **g.** Leaf margin. **h.** Basal marginal leaf cells. **i.** Habit. **j.** Portion of stem. **k.** Leaves, showing variation in shape. **l.** Leaf apex. **m.** Leaf margin. **n.** Basal marginal leaf cells. Reproduced with permission from Mem. New York Bot. Gard. 82: plate 141, figs. 1–14. 1998.

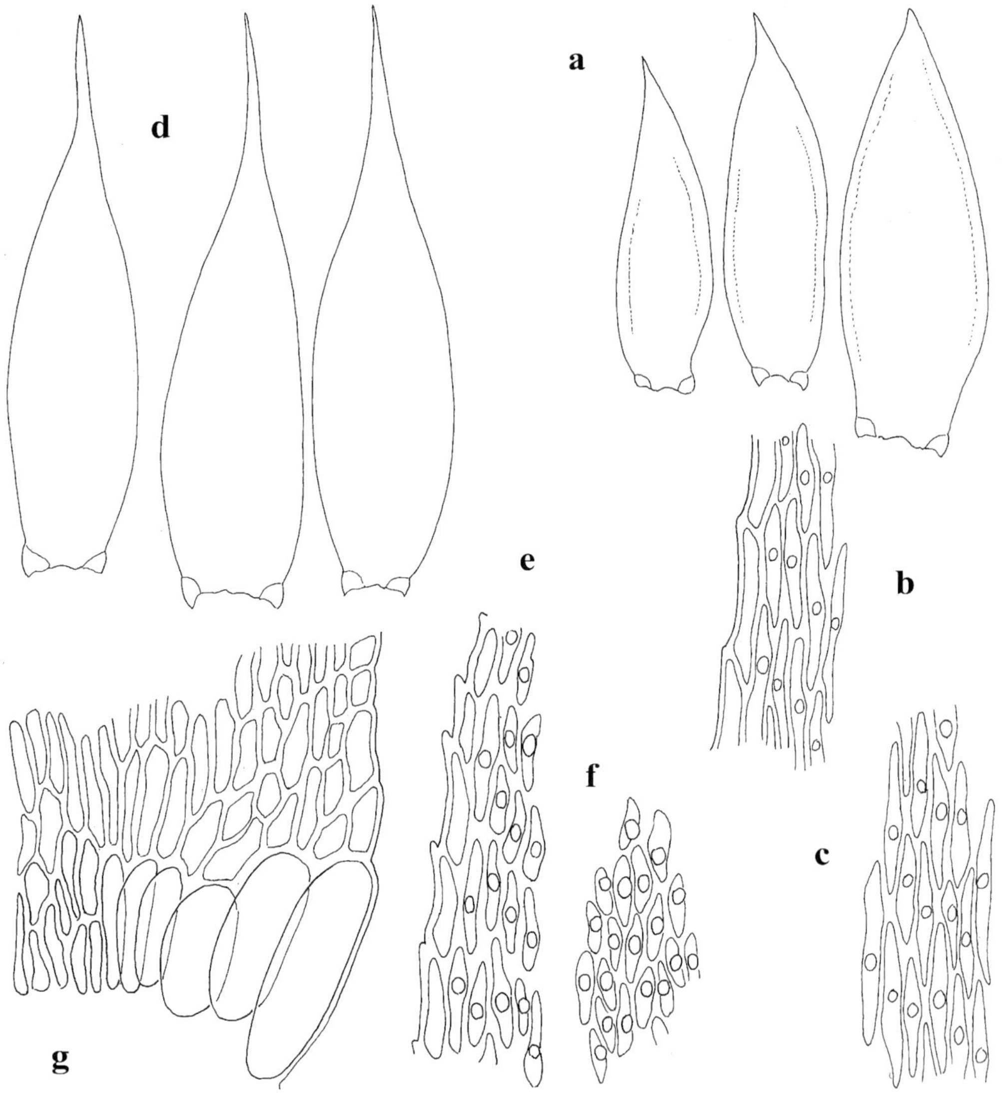

FIG. 55. *Trichosteleum fluviale.* **a.** Leaves, x 95. **b.** Upper marginal cells of leaf, x 468. **c.** Upper median cells of leaf, x 468. **d.** Leaves, x 95. **e.** Upper marginal cells of leaf, x 468. **f.** Upper median cells of leaf, x468. **g.** Algar cells of leaf, x 468. Reproduced with permission from Mem. New York Bot. Gard. 69(2): figure 750. 1994.

SPHAGNACEAE

SPHAGNUM L.

Plants small to robust, whitish green to yellowish brown, sometimes tinged with purple or reddish color, in mats, cushions or hummocks, growing in wet habitats or in peat bogs. **Stems** erect, simple or sparsely branched; the branches often arranged in fascicles, usually crowded at the stem tip forming a capitulum. **Stem leaves** usually differentiated in size and shape from branch leaves. **Branch leaves** often bordered by several rows of linear cells, enclosing a network of elongated green cells surrounding large, empty hyaline cells; the hyaline cells nearly always reinforced by ringlike fibrils and porous at the surfaces. **Sporophytes** consisting of a capsule and a foot, nearly sessile, elevated at maturity by elongation of the gametophytic stalk-like pseudopodium; **capsules** globose, dark brown to black; opercula developed; annulus none; peristome absent. **Calyptrae** delicate, with hyaline membrane covering the opercula and irregularly ruptured when capsules mature. **Spores** large, tetrahedral.

58. Sphagnum subsecundum Nees ex Sturm (**Fig. 56**)

Plants small to robust, grayish green to often yellow or brown, in loose cushions. **Stems** up to 20 cm high; cortex in 1 layer, hyaline cells without fibrils and pores; branches in fascicles of 3–5. **Stem leaves** 0.5–1.0 mm × 0.4–0.8 mm, triangular-ligulate or ligulate, rounded, often hyaline and lacerate at the apex. **Branch leaves** 1.0–1.5 mm × 0.5–0.6 mm, broadly ovate-lanceolate, strongly concave, blunt and dentate at the apex; hyaline cells fibrillose, with a single, small pore at the corners or numerous small, ringed pores along commisural rows on the dorsal surface; chlorocysts in cross-section narrowly rectangular or narrow elliptic, centrally located and exposed on both surfaces. **Spores** yellow, papillose, 25–28 µm diameter.

Notes.—Reported by Vargas (1974). Unlikely to be found in the lowland area of Madre de Dios, unless in *Mauritia flexuosa* palm swamps.

Habitat.—Open montane forest, 1400–3500 m, on sandy soil over rocks, along streams.

SPLACHNOBRYACEAE

SPLACHNOBRYUM Müll. Hal.

Plants small. **Stems** infrequently branched. **Leaves** more or less crispate when dry, erect-spreading when moist, ovate, ligulate, rarely lanceolate; apices rounded to obtuse; margins bordered or not, plane to recurved below, crenulate at apices, entire below; **costa** single, ending far below apices to ± percurrent; cells unistratose, thin-walled, smooth, upper cells rhombic, basal cells oblong. **Setae** terminal or lateral, elongate; **capsules** ovoid to cylindrical; annulus none; opercula conical; peristome single. **Calyptrae** cucullate. **Spores** finely papillose or smooth.

59. Splachnobryum obtusum (Brid.) Müll. Hal. (**Fig. 57**)

Splachnobryum spruceanum Müll. Hal.

Stems to 12 mm long. **Leaves** 1–2 mm long, ovate to ligulate; apices obtuse to rounded; margins entire to crenulate, particularly in distal half; **costa** subpercurrent to percurrent; cells smooth, occasionally bulging-mammillose, upper cells rhombic 1–2:1, median cells 2–4:1, basal cells oblong. **Setae** yellow to reddish brown, 3–6 mm long; **capsules** 0.6–2 mm long. **Spores** smooth, 13–17 µm diameter.

Habitat.—Lowland lower montane forest, sea level to 850 m; on wet rocks, shaded soils and in seepage areas.

STEREOPHYLLACEAE

Plants medium-sized to moderately robust, in often lustrous, green to golden, thin to dense, flat mats. **Stems** creeping, simple or sparingly branched, terete- or complanate-foliate. **Stem and branch leaves** similar, crowded or sometimes lax, usually imbricate when dry, erect to wide-spreading when moist, lanceolate to ovate, obtuse to long-acuminate, symmetrical or asymmetrical, flat to somewhat concave, not plicate, not decurrent; margins usually serrulate above, ± entire below, mostly plane; **costa** mostly single, ending 1/3–3/4 the leaf length, rarely weaker or absent in some leaves, occasionally projecting as a spine or prickle at apex; cells rhomboidal to linear, smooth or rarely unipapillose or prorulose at upper ends at back, thin- to thick-walled; alar cells differentiated in mostly large areas, oblate or quadrate to rectangular, often unequally distributed on either side of the costa but always covering the adaxial surface of the costa. **Setae** elongate, straight; **capsules** erect to cernuous, straight to curved, ovoid to cylindrical; annulus differentiated and deciduous or undifferentiated; opercula conical to short-rostrate; peristome double, exostome shouldered, bordered, with a zig-zag median line on the front surface, cross-striolate or with horizontal arranged papillae below, papillose above. **Calyptra** cucullate, naked, smooth. **Spores** spherical to ovoid, papillose.

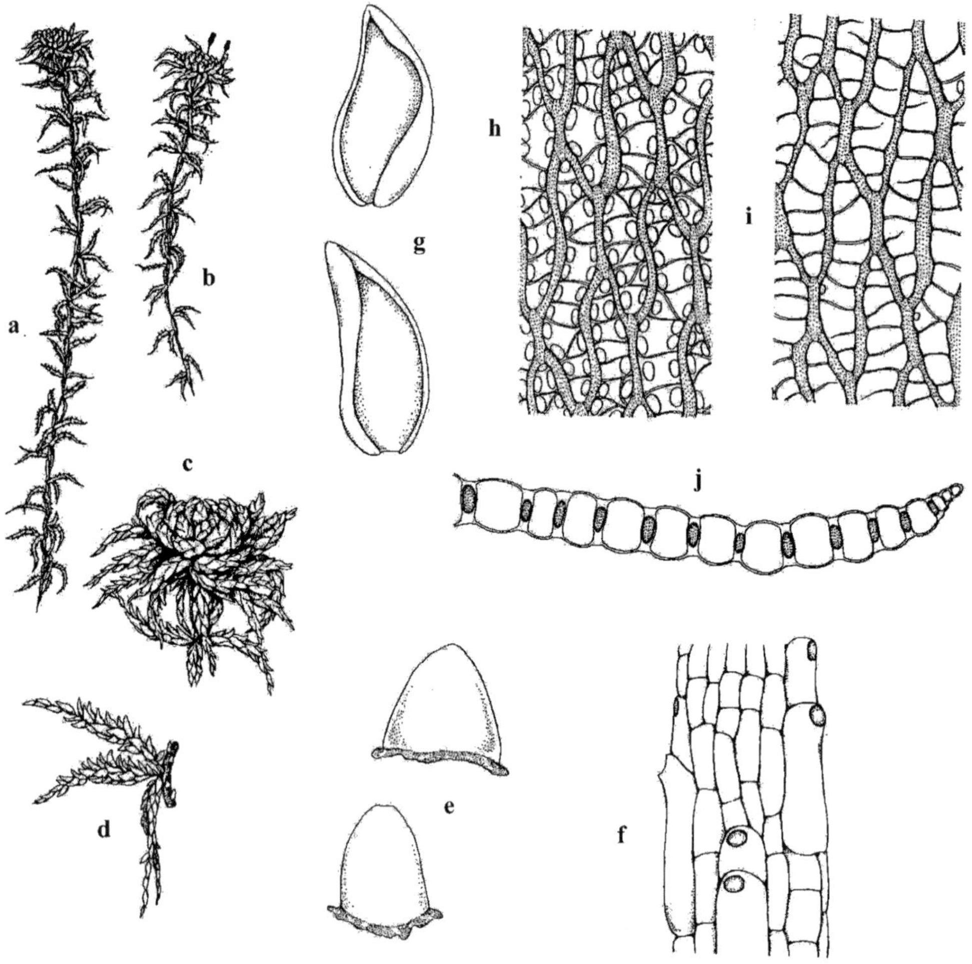

Fig. 56. *Sphagnum subsecundum*. **a.** Habit, dry. **b.** Habit with capsules, dry. **c.** capitulum, dry. **d.** fascicle, dry. **e.** Stem leaves. **f.** Branch cortex in surface view. **g.** Branch leaves. **h.** Upper cells of branch leaves, outer surface. **i.** upper cells of branch leaf inner surface. **j.** Branch leaf cross section. Reproduced with permission from Crum and Anderson (1981).

PILOSIUM (Müll. Hal.) M. Fleisch.

Plants medium-sized, in lustrous, yellow-green, mostly dense, ± flat, often extensive mats. **Stems** creeping, irregularly branched, complanate-foliate. **Stem and branch leaves** similar but **lateral and dorsal leaves** differentiated, crowded, ± stiff, wide-spreading, somewhat curled when dry, curled downward when moist, complanate, **dorsal leaves** oblong-ovate, symmetrical, **lateral leaves** oblong-lanceolate, cultriform, obtuse to short-acute, flat or concave, not plicate, not decurrent; margins entire to serrulate above, entire below, plane; **costa** single in lateral leaves, weak, rarely extending 1/3 the leaf length, absent in dorsal leaves; cells linear, smooth, thin- to firm-walled, porose or not, heavily chlorophyllose; alar cells differentiated in lateral leaves, numerous, large, oblong, colored, porose, extending across adaxial surface of costa, dorsal leaves with few undifferentiated cells, small, quadrate to subquadrate. **Setae** elongate, orange-red, straight; **capsules** cernuous, symmetrical, not curved, short-cylindrical, smooth, constricted below the mouth when dry; opercula short-rostrate; annulus not differentiated; exostome teeth cross-striolate below, papillose above. **Spores** spherical.

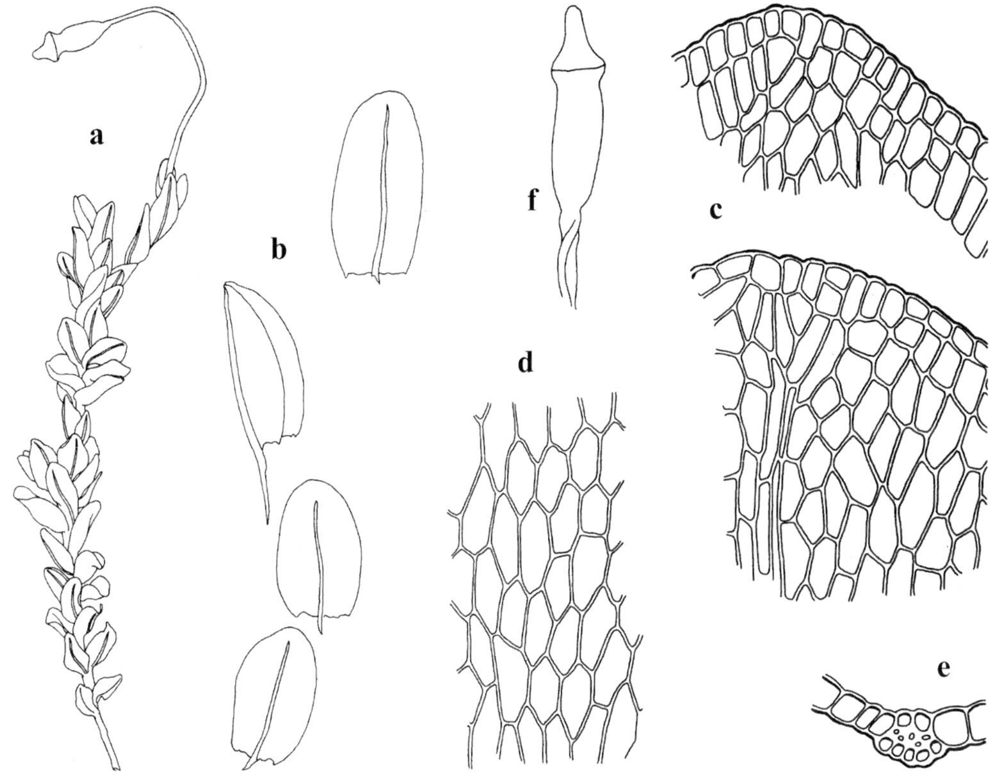

Fig. 57. *Splachnobryum obtusum*. **a.** Habit, × 9. **b.** Leaves, × 30. **c.** Apical cells of leaf, × 337. **d.** Median cells of leaf, × 337. **e.** Costa in section, × 337. **f.** Capsule, × 30. Reproduced with permission from Mem. New York Bot. Gard. 69(1): fig. 333. 1994.

60. Pilosium chlorophyllum (Hornsch.) Müll. Hal. (**Fig. 58**)

Stems creeping, to 6 cm long. **Leaves** differentiated, **dorsal leaves** erect, oblong-ovate, 1.3–1.6 mm long, **lateral leaves** wide-spreading, oblong-lanceolate, 1.5–2 mm long, cultriform, obtuse to short-acute; cells linear, 90–150 µm long, porose or not; alar cells strongly differentiated in lateral leaves, numerous, 25–90 × 14–28 µm, oblong, often colored, ± inflated, porose, dorsal leaves with few differentiated cells, 12–24 × 9–12 µm, quadrate to subquadrate. **Setae** 1.5–1.8 cm long; **capsules** short-cylindrical, orange-red, 0.8–1.2 mm long. **Spores** 7–9 µm diameter.

Habitat.—Wet lowland to lower montane forest, 0–1000 m, on logs, less often on tree bases, litter or rocks. Widespread in the neotropical lowlands.

THAMNOBRYACEAE

Plants mostly medium-sized to large and robust, forming loose to dense tufts or mats. **Primary stems** creeping; **secondary stems** usually stiffly erect and frondose; stipitate or indistinctly so, irregularly to regularly pinnately branched, distal branch tips often flagelliform. **Primary stem leaves** mostly scale-like; **stipe leaves** triangular to ovate-triangular. **Secondary stem leaves** ovate-oblong to oblong-lanceolate or oblong-ligulate, concave, smooth to weakly plicate, apex acute to obtuse, base weakly auriculate or not; margins plane to recurved (usually below), serrate distally, often coarsely and irregularly so at apex; **costa** single, occasionally forked, 1/4–4/5 lamina length, usually strong; apical cells short, median cells linear to fusiform or rhomboidal, walls entire or weakly porose, smooth to occasionally papillose by projecting cell

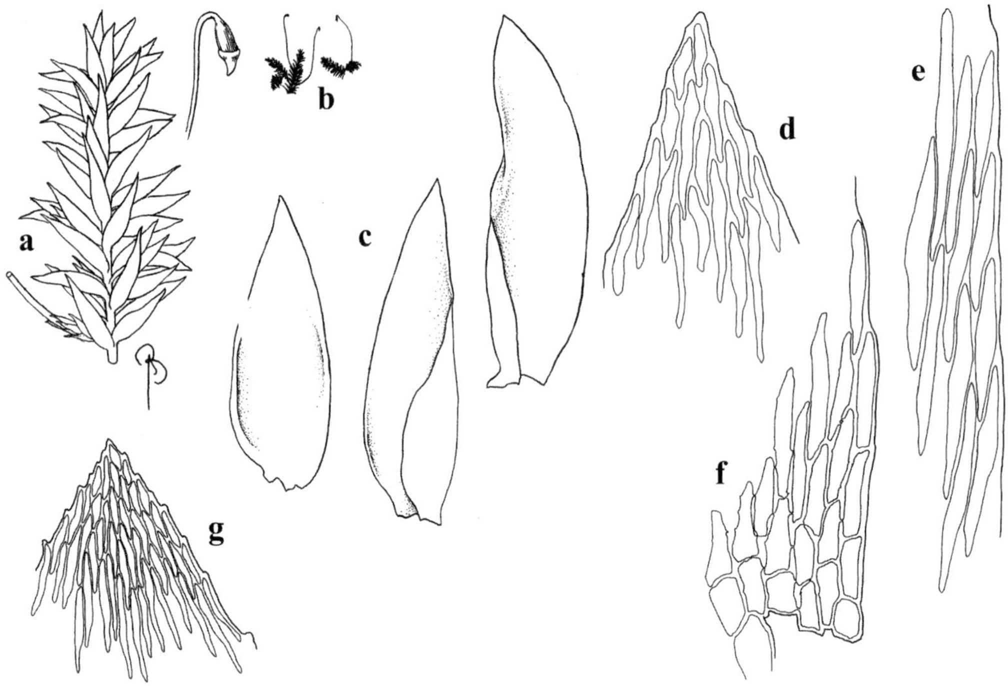

Fig. 58. *Pilosium chlorophyllum*. **a.** Portion of stem and sporophyte. **b.** Habit. **c.** Leaves. **d.** Leaf apex. **e.** Leaf margin. **f.** Basal marginal leaf cells. **g.** Leaf apex of *Pilosium serrulatum*. Reproduced with permission from Mem. New York Bot. Gard. 82: plate 111, figs. 1–7. 1998.

angles, alar region differentiated or not. **Branch leaves** often differentiated, usually narrower, oblong-ligulate; flagelliform leaves reduced, scale-like. **Seta** elongate; **capsules** erect to horizontal, cylindrical to ovoid or obloid; opercula short- to long-rostrate; peristome double, exostome teeth 16, cross-striolate or striate-papillose below, distally papillose. **Calyptrae** cucullate, smooth, and naked. **Spores** spherical, papillose.

THAMNOBRYUM Nieuwl.

Plants medium-sized or more often robust, stipitate-frondose, erect to sometimes pendent, in slightly lustrous to dull, dark- to yellow-green, often extensive colonies. **Secondary stems** arising from upturning primary stems, the creeping stem continuing by innovations near the base of the stipe; secondary stems distinctly stipitate, sparsely to densely, irregularly to pinnately branched, ± complanate-foliate. **Primary stem and stipe leaves** similar, but those of stipe larger, appressed with spreading apices on stipe, not complanate, ± triangular, decurrent or not; margins mostly subentire; **costa** single but mostly obscure, sometimes absent; cells mostly elongate, smooth, thick-walled; alar cells not or scarcely differentiated. **Secondary stem and branch leaves** similar, those of branches sometimes smaller, ± complanate, mostly spreading, ovate, ovate-lanceolate, oblong-ovate or oblong-ligulate, obtuse and mucronate to short-acuminate, sometimes plicate; margins serrate to serrulate at apex, subentire to serrulate below, plane or narrowly recurved; **costa** single, strong, ending ca. 4/5 the leaf length to subpercurrent; cells mostly long-hexagonal at midleaf, smooth or rarely prorulose, ± porose, becoming shorter at the apex, rhomboidal to hexagonal, usually becoming longer toward the insertion; alar cells not or poorly differentiated. **Setae** elongate; **capsules** inclined to horizontal to pendent, cylindrical; opercula obliquely rostrate from a conical base; annulus differentiated; exostome teeth cross-striolate below, papillose above. **Spores** finely papillose to apparently smooth.

61. Thamnobryum fasciculatum (Hedw.) I. Sastre (**Fig. 59**)

Plants robust, stipitate-frondose. **Primary stems** creeping, **secondary stems** arising from upturning primary stems, 5–20 cm tall, erect to pendent, pinnately to irregularly branched, ± complanate-foliate. **Primary stem leaves** widely spaced, appressed with wide-spreading to squarrose apices, not complanate, broadly triangular, to ca. 1.3 mm long, narrowly acuminate, ± channeled above, narrowly decurrent; margins subentire throughout except for a few small teeth at extreme base, sometimes serrulate at base of acumen; **costa** diffuse, ± single and very short or short and double to absent; cells rounded linear-rectangular, ca. 10–15:1, smooth, thick-walled, porose, becoming short across the insertion; alar cells little differentiated. **Stipe leaves** appressed with erect-spreading apices, not complanate, ovate-triangular, 2.1–3 mm long, short- to long-acuminate, ± channeled above, decurrent; margins subentire above, serrulate below; **costa** single, ending 1/3–2/3 the leaf length; cells linear 12–20:1, smooth or prorulose, porose, becoming shorter toward the insertion; alar cells not or scarcely differentiated. **Secondary stem and branch leaves** similar, erect-spreading, ovate-ligulate, 2.2–4 mm long, obtuse and apiculate to acute, typically plicate, especially when dry; margins serrate above, serrulate to base, plane or narrowly recurved near midleaf; **costa** single, strong, subpercurrent, not projecting at apex; cells long-hexagonal at midleaf, ca. 6–8:1, smooth to prorulose, not or slightly porose, upper leaf cells becoming conspicuously shorter, ca. 1–2:1, mostly ± hexagonal to sub-quadrate to rhombic, becoming long-rectangular and porose toward the insertion; alar cells not or scarcely differentiated. **Setae** elongate, red-brown, 1.5–2.5 cm long; **capsules** horizontal to pendent, cylindrical, 1.6–2.8 mm long. **Spores** ± smooth, 10–14 µm diameter.

Habitat.—Lower montane to midmontane forest, 900–2700 m, in moist, shaded habitats, especially on rocks and soil near waterfalls and streams. Recorded by Vargas (1974) but not likely in the lowlands of Madre de Dios.

THUIDIACEAE

Plants small to robust, in typically stiff, dark-green to golden, loose mats. **Stems** creeping or ascending, mostly regularly 1–3 pinnate, not at all stipitate. **Stem and branch leaves** strongly differentiated. **Stems leaves** mostly spreading when moist, usually broadly ovate, often abruptly acuminate, concave; margins serrulate to crenulate ± throughout, plane or recurved; **costa** single, strong, ending in the acumen to ± excurrent; cells quadrate to short-rectangular, uni- or pluripapillose on both surfaces or just at back, firm-walled; alar cells not or poorly differentiated. **Branch leaves** smaller than stem leaves or not, usually spreading, lanceolate to ovate, usually gradually acuminate to obtuse; margins serrulate to crenulate ± throughout, usually plane; **costa** single, weaker than in stem leaves; cells ± quadrate, uni- or pluripapillose on both surfaces or just at back, the apical cell typically truncate and pluripapillose, firm-walled, not porose; alar cells not or poorly differentiated. **Setae** elongate, smooth to roughened, reddish; **capsules** mostly inclined to horizontal, rarely erect, mostly asymmetrical and curved, cylindrical; opercula usually rostrate; annulus differentiated; peristome double, exostome teeth shouldered, bordered, cross-striolate below on the front surface, sometimes with overlying papillae, papillose above. **Calyptrae** cucullate, rarely mitrate, mostly naked, smooth or roughened. **Spores** spherical, usually papillose.

CYRTO-HYPNUM (Hampe) Hampe & Lorentz

Plants small and slender, mostly green but occasionally golden, lax mats. **Stems** creeping, 1–2(–3) pinnate, not frondose; paraphyllia few to relatively abundant, usually confined to the stems, filamentous, mostly unbranched. **Stem and branch leaves** little differentiated. **Leaves** mostly broadly ovate, acuminate, seldom plicate, mostly short-decurrent or subauriculate; margins mostly crenulate-papillose, less often serrate to serrulate, often recurved at least below; **costa** single, usually strong but not filling the acumen (weaker in branch leaves); cells isodiametric to quadrate or short-rectangular, mostly pluripapillose, occasionally uni-papillose, always on both surfaces, thin- to firm-walled, sometimes becoming more elongate in the acumen; alar cells not differentiated. **Setae** elongate, smooth to roughened or papillose above or throughout, reddish,

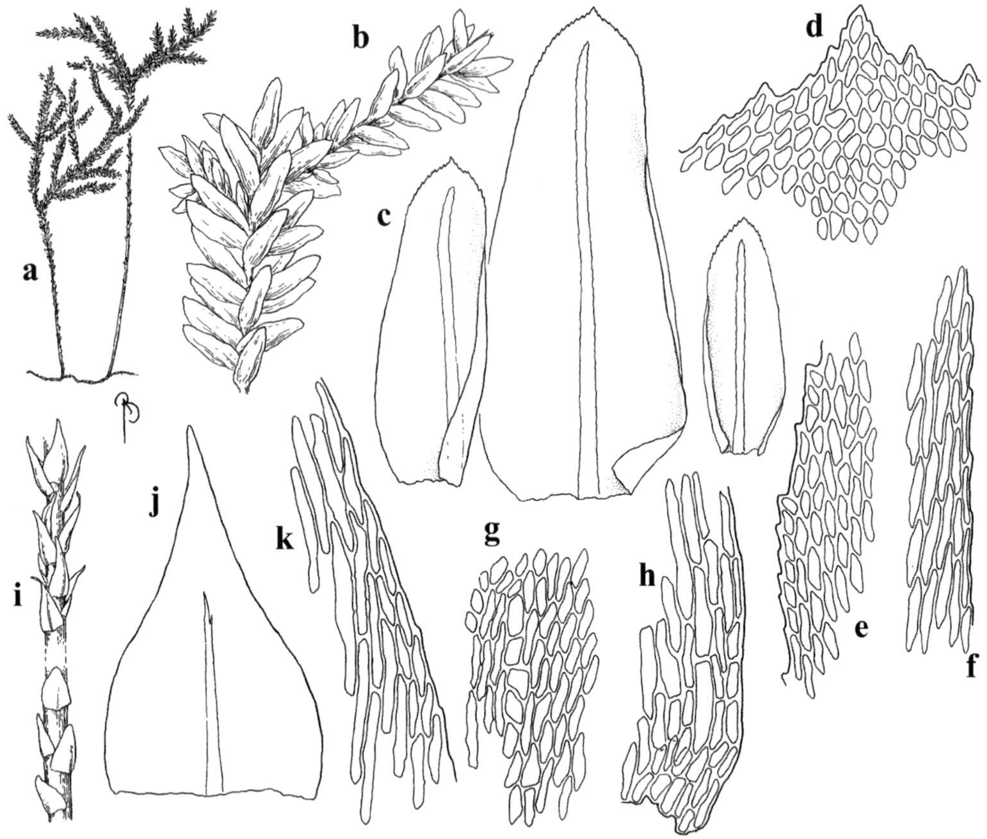

Fig. 59. ***Thamnobryum fasciculatum*. a.** Habit. **b.** Portion of secondary stem. **c.** Secondary stem and primary branch leaves. **d.** Secondary stem leaf apex. **e.** Secondary stem leaf margin. **f.** Primary branch leaf margin. **g.** Juxtacostal laminal cells. **h.** Basal marginal secondary stem leaf cells. **i.** Portion of stipe. **j.** Stipe leaf. **k.** Stipe leaf margin. Reproduced with permission from Mem. New York Bot. Gard. 82: plate 46, figs. 1–11. 1998.

capsules erect to horizontal, rarely ± pendent, often asymmetrical and curved, cylindrical; opercula conical, often obliquely rostrate. **Calyptrae** cucullate. **Spores** usually finely papillose.

1. Plants 1-pinnate; branch leaves strongly incurved when dry ________ **Cyrto-hypnum involvens**
1. Plants 2-pinnate; branch leaves not or only slightly incurved when dry ________ **Cyrto-hypnum scabrosulum**

62. Cyrto-hypnum involvens (Hedw.) W.R. Buck & H.A. Crum (Fig. 60)

Plants small and slender, in typically stiff, dull, mostly dark-green, often dense mats. **Stems** creeping, to ca. 3 cm long, once-pinnate; paraphyllia usually very sparse on stems, absent on branches, filamentous, 1(–2)-seriate, unbranched, to ca. 8 cells long, the cells quadrate, the distal cell truncate. **Stem and branch leaves** scarcely differentiated, stem leaves flexuosely incurved when dry, spreading when moist, ovate-triangular, 0.3–0.6 mm long, ± abruptly short-acuminate, concave, not plicate, not decurrent; margins crenulate-papillose throughout, plane or narrowly recurved below; **costa** single, strong, disappearing in the acumen but not filling it; cells ± quadrate, obscurely pluripapillose. **Branch leaves** strongly incurved when dry, spreading when moist, oblong to oblong-ovate, ca. 0.5 mm long, obtuse to acute, concave, not plicate, not decurrent; margins regularly crenulate-serrulate, plane or narrowly recurved below; **costa** single, ending ca. 7/8 the leaf length, often ± flexuose, apex projecting at back as a small spine; cells ± quadrate, pluripapillose. **Setae** elongate, roughened throughout, orange, 0.8–1.5 cm long; **capsules** inclined to subpendent, broadly short-cylindrical,

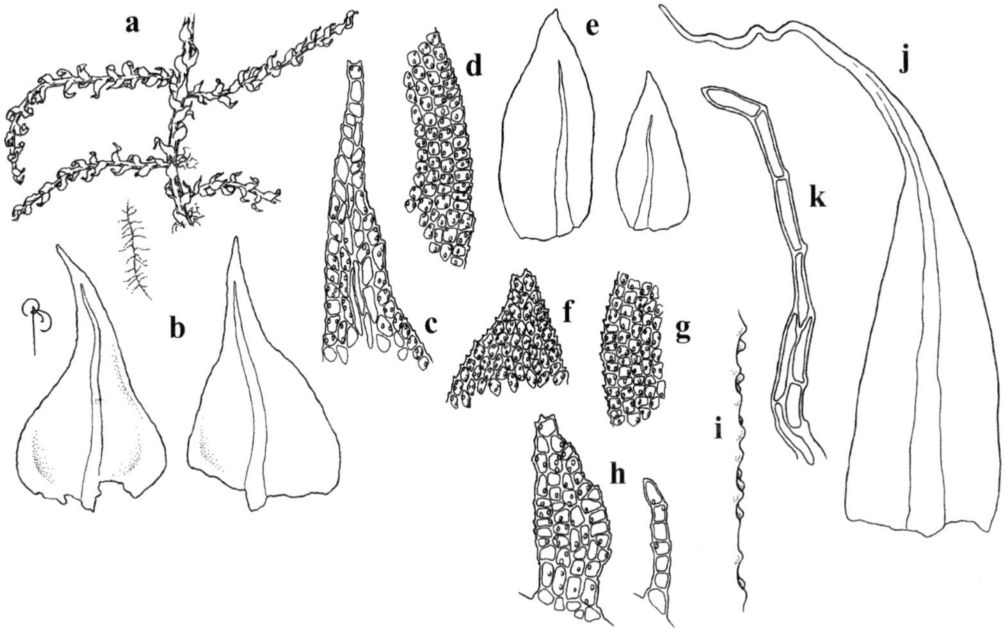

Fig. 60. *Cyrto-hypnum involvens.* **a.** Habit and portion of stem with branches. **b.** Stem leaves. **c.** Stem leaf apex. **d.** Stem leaf margin. **e.** Branch leaves. **f.** Branch leaf apex. **g.** Branch leaf margin. **h.** Pseudoparaphyllium (left) and paraphyllium (right). **i.** Profile of seta. **j.** Perichaetial leaf. **k.** Perichaetial leaf apex. Reproduced with permission from Mem. New York Bot. Gard. 82: plate 75, figs. 1–11. 1998.

1–1.3 mm long; opercula obliquely long-rostrate; exostome teeth yellow to yellow-brown, shouldered, bordered, cross-striolate below on the front surface, becoming papillose above. **Spores** 11–14 µm diameter.

Habitat.—Dry to mesic, less often humid forest, ca. 100–1000 m, on soil, limestone, humus and old wood.

63. Cyrto-hypnum scabrosulum (Mitt.) W.R. Buck & H.A. Crum (Fig. 61)

Plants small and slender, in typically stiff, dull, mostly green to yellow-green, often lax mats. **Stems** creeping, to ca. 6 cm long, regularly 2-pinnate; paraphyllia moderately abundant on stems, sparse at base of primary branches, absent on secondary branches and distal end of primary branches, filamentous, uniseriate, unbranched, 2–5 cells long, the distal cell truncate. **Stem and branch leaves** strongly differentiated. **Stem leaves** appressed when dry, erect-spreading when moist, broadly ovate-triangular, 0.3–0.6 mm long, ± abruptly broadly acuminate, concave, not plicate, short decurrent; margins crenulate-papillose throughout, narrowly recurved below; **costa** single, broad, ending in but not filling the acumen; cells short-rectangular, pluripapillose. **Branch leaves** incurved when dry, spreading when moist, laxly disposed so that branches obviously exposed when dry, ovate, 0.15–0.25 mm long, obtuse to broadly acute, concave, not plicate, subauriculate; margins regularly crenulate-papillose, plane; **costa** single, ending ca. 2/3 the leaf length, not or scarcely projecting at back; cells ± quadrate, pluripapillose, ca. 6 µm wide. **Setae** elongate, roughened throughout, orange, 0.7–2 cm long; **capsules** horizontal to pendent, short-cylindrical, 0.5–1 mm long; opercula obliquely long-rostrate; annulus of 1–2(–3) rows of small, rectangular, firm-walled cells; exostome teeth yellow to yellow-brown, shouldered, bordered, on the front surface cross-striolate below, becoming papillose and then coarsely papillose above. **Spores** 14–18 µm diameter.

Notes.—The report of *C. leptocladum* (Majestyk & Janovec 2004) was erroneous and can be referred to *C. scabrosulum*.

Habitat.—Mostly moist lowland, to ca. 300 m, on limestone and old wood.

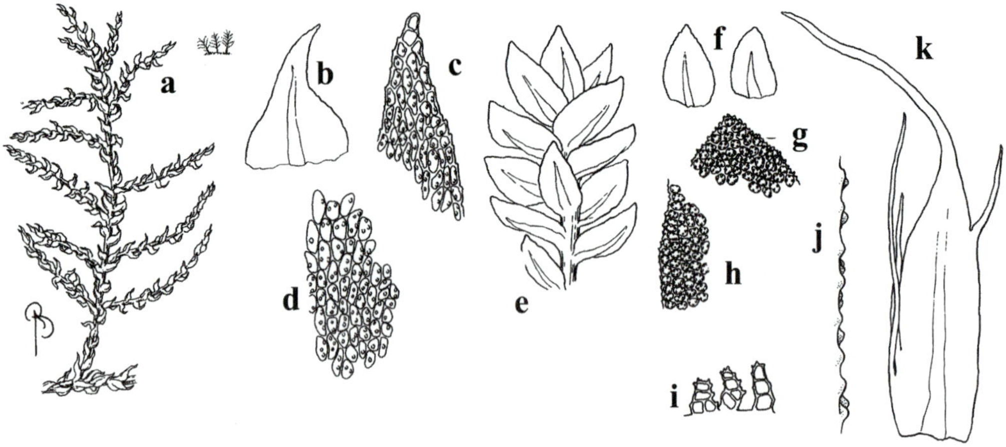

Fig. 61. *Cyrto-hypnum scabrosulum*. **a.** Habit and portion of stem with branches. **b.** Stem leaf. **c.** Stem leaf apex. **d.** Laminal cells, midleaf. **e.** Portion of branch. **f.** Branch leaves. **g.** Branch leaf apex. **h.** Branch leaf margin. **i.** Paraphyllia. **j.** Profile of seta. **k.** Perichaetial leaf. Reproduced with permission from Mem. New York Bot. Gard. 82: plate 76, figs. 1–11. 1998.

Fig. 62. *Calymperes afzelii*. Photographed by John Janovec and Piers Majestyk.

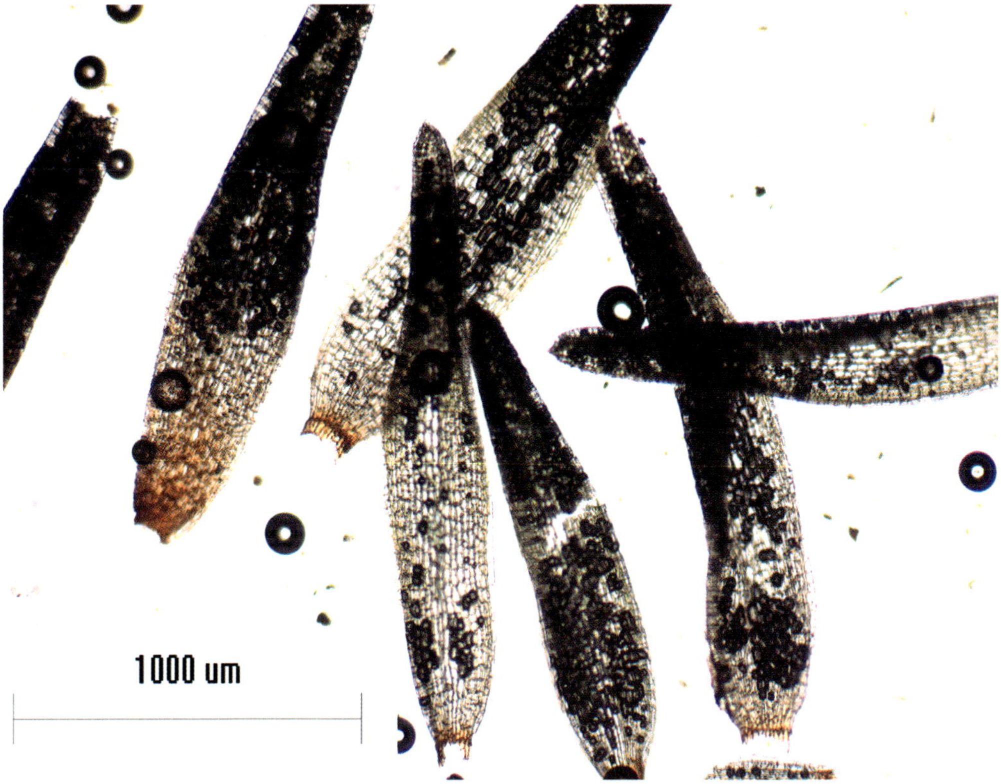

Fig. 63 a. *Leucobryum subobtusifolium*. Photographed by John Janovec and Piers Majestyk.

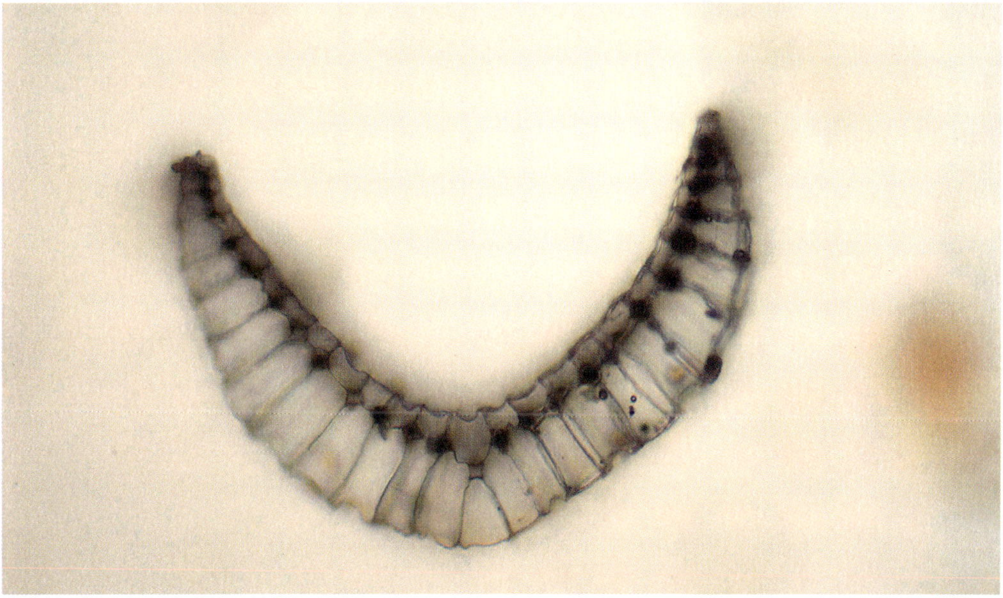

Fig. 63 b. *Leucobryum subobtusifolium*. Photographed by John Janovec and Piers Majestyk.

FIG. 64. *Octoblepharum cylindricum*. Photographed by John Janovec and Piers Majestyk.

FIG. 65. *Octoblepharum pulvinatum*. Photographed by John Janovec and Piers Majestyk.

Fig. 66. *Henicodium geniculatum*. Photographed by John Janovec and Piers Majestyk.

GLOSSARY

Abaxial Side away from stem or axis; back, dorsal, or lower surface of leaf or costa.

Acrocarpous With gametophyte producing sporophyte at apex of a stem or main branch. Acrocarpous mosses generally grow erect in tufts (rather than mats) and are sparsely or not branched (see **Pleurocarpous**).

Acumen A slender, tapering point.

Acuminate Slenderly tapered with an angle of less than 45°; longer than acute.

Acute Sharp pointed, with terminal angle less than 90° but greater than 45°.

Adnate Fused together; the fusion of unlike parts, i.e. perianth and bract (see **Connate, Adherent**).

Alar cells Referring to cells at basal margins (angles) of a leaf; these cells are often differentiated in size, shape, or color from other leaf cells.

Anisosporous Having spores with more than a single distribution of size, generally the smaller spores giving rise to male gametophytes.

Annulus In stegocarpous mosses a zone of variously differentiated cells between the capsule urn and operculum, facilitating opening of the capsule.

Antheridium Male gametangium; a multicellular globose to broadly cylindrical, stalked structure containing spermatozoids.

Apiculate Abruptly short-pointed (**Mucronate** is shorter; **Cuspidate** is longer and stouter).

Apiculus A short, abrupt point.

Apophysis A strongly differentiated, sterile neck at the base of a capsule, between seta and urn.

Appressed Closely applied, as leaves lying closely or flat against the stem.

Archegonia Female gametangium or sex organ; a multicellular, flask-shaped structure consisting of a stalk, venter, neck, and containing an ovum.

Areolation The cellular network of a leaf or thallus.

Ascending Pointing obliquely upward; away from the substrate.

Asymmetrical Not symmetrical; not capable of division by a longitudinal plane into similar halves.

Auricle A small, ear-like lobe; often present at the basal margins of a leaf in mosses.

Auriculate With auricles.

Autoicous With archegonia and antheridia in separate clusters (gametoecium) on the same plant.

Awn Hair-like point, usually formed by an excurrent costa.

Axillary In the leaf axils.

Basal membrane A short tube or cylinder often supporting segments and cilia of the endostome.

Bifid Divided or branched into two parts.

Bistratose Composed of two cell layers; e.g., leaf blades that are two cells thick.

Branch A lateral division of the stem, or axis of growth.

Brood body A generalized term used to denote various types of specialized vegetative reproductive structures; e.g., reduced buds, leaves, branches, or plant fragments (propagules).

Caespitose Tufted; growing in cushions or sods.

Calyptra A membranous covering of haploid tissue over the developing sporophyte, derived largely from the archegonial venter. In mosses the venter generally ruptures near the base, is carried upward by elongation of the seta, and frequently expands to form a protective covering over the capsule.

Campanulate Bell-shaped, referring to a calyptra that is elongated and cylindrical.

Cancellina Large, empty, and usually hyaline basal leaf cells (hyalocysts); particularly those found in leaves of pottioid families such as Calymperaceae and Encalyptaceae, where the hyalocysts form a well-marked lattice-like region.

Capitulum Head.

Capsule The sporangium; terminal spore-producing part of the sporophyte; in most mosses it is differentiated into an apical operculum, central urn (spore-bearing region), and a sterile basal neck or hypophysis.

Carinate With a keel or ridge.

Chlorocysts Green (chlorophyllose) cell; generally used in contrast to hyalocysts; e.g., leaves of *Sphagnum*, *Leucobryum* or *Calymperes*.
Ciliate Fringed with hair-like appendages.
Clavate Thickened towards the apex; club-shaped.
Complanate Flattened or compressed, such as leaves flattened into more or less one plane.
Concolorous Uniform in color.
Conduplicate Strongly folded longitudinally along the middle.
Confluent Blending of two structures into one so as to obscure the evidence of union.
Conical Cone-shaped; e.g., operculum of *Bryum*.
Connivent Coming into contact but not joined.
Contorted Irregularly curved or twisted.
Convex Curving or bulging outward.
Convolute Rolled together and forming a sheath.
Corticolous Growing on bark.
Costa Nerve or midrib of a leaf, always more than one cell thick.
Crenulate With minute, rounded teeth.
Crispate Wavy or notched and curled irregularly.
Crisped Wavy; often used more loosely to mean variously curled, twisted, and contorted.
Cucullate Hooded or hood-shaped; a calyptra split along one side only; also used to describe leaves strongly concave and erect or inflexed at the tips, like a monk's cowl.
Cultriform Shaped like a knife or plowshare; curved and asymmetrical.
Cuspidate Ending abruptly in a stout, rigid point.
Cylindrical Elongate and circular in transverse section.
Cymbiform Concave and broadly boat-shaped.
Deciduous Falling off; e.g., shedding of operculum.
Decurrent With basal leaf margins extending down the stem past the leaf insertion as ridges or narrow wings.
Dehiscent The capsule opening regularly by means of an annulus and operculum or valves.
Deltoid Shaped like an equilateral triangle.
Denticulate Finely toothed.
Depauperate Poorly developed.
Dimorphic Occurring in two forms; e.g., with leaves of two forms.
Dioicous With archegonia and antheridia on separate plants.
Diplolepidous Form of arthrodontous peristome; (**originally**) having main teeth with two columns of cells up dorsal face; (**strictly**) having two concentric circles of teeth, with the outer circle (exostome) derived from thickening of the contiguous walls of the outer and primary peristomial layers and the inner circle (endostome) derived from thickening of the contiguous walls of the primary and inner peristomial layers. The exostome is generally more heavily thickened than the endostome; however one or both circles may be vestigial.
Distally Away from the base or point of attachment; towards the apex of a leaf or stem; the outer, convex face of a spore.
Distichous Leaves alternating in two opposite rows; e.g., *Fissidens*.
Dorsal (Of leaves) the abaxial, back, or lower surface; (of peristome teeth) the outer face; (of stems or thalli) the upper surface, away from the substrate.
Elimbate Lacking a border; referring to leaves without strongly differentiated marginal cells.
Emarginate Broad, shallowly notched apex, deeper than retuse.
Endostome The inner circle of a diplolepidous (double) peristome, formed from contiguous periclinal wall-pairs of the primary and inner peristomial layers; typically a weak membranous structure consisting of a basal membrane bearing segments and cilia. The endostome is homologous with the haplolepidous (single) peristome.
Entire Without teeth, more or less smooth on the margin; e.g., leaves.

Epiphyllous Growing on leaves of other plants.

Erect With leaves directed toward stem apex; with leaf margins curved upward; with capsules straight, not curved.

Excavate Hollowed out.

Excurrent Extending beyond the apical margin; e.g., an awn formed by a protruding costa.

Exerted Protruding beyond another structure.

Exostome The outer circle of the diplolepidous (double) peristome, formed from contiguous periclinal wall-pairs of the outer and primary peristomial layers; missing or rudimentary in haplolepidious (single) peristomes.

Falcate Curved like the blade of a sickle.

Fasciculate Bunched together, in bundles or fascicles; e.g., branches of *Sphagnum*.

Fastigiate With branches erect and more or less equal in length.

Fibrillose With fine, fiber-like wall thickenings (fibrils); applied to *Sphagnum* hyalocysts where the fibrils may be spiral or annular.

Fibril Fine, fiber-like wall thickenings.

Filiform Slender and elongate, filamentous, thread-like.

Flaccid Soft, limp.

Flagellate Whiplike; branches with a gradual attenuation from ordinary leaves at the branch base to vestigial-leaved branch tips.

Flexuose Slightly and irregularly bent, twisted, or wavy.

Foot The basal portion of most bryophyte sporophytes, embedded in the gametophyte and serving as both an organ of absorption and attachment.

Fringed With a shortly ciliate margin or edge.

Frondose Resembling a fern frond; (in mosses-) closely and regularly branched in one plane (pinnate); e.g., *Thuidium*.

Furrowed Grooved.

Fusiform Spindle-shaped; narrow (more than 3 times as long as wide) and tapered at both ends.

Gemmae Uni- or multicellular, filamentous, globose, ellipsoidal, cylindrical, stellate or discoid brood bodies, relatively undifferentiated, serving in vegetative reproduction.

Gemmiferous Bearing gemmae.

Glaucous With a whitish, greyish, or bluish overcast, hue or color.

Globose Spherical.

Granular Roughened with minute, blunt projections.

Gregarious Growing together in loose tufts or mixed mats.

Guttulate With cell lumen rounded because of strong wall thickening; resembling drops of oil or resin.

Gynoecium The female gametoecium, consisting of archegonia and the surrounding bracts (perichaetial leaf or involucre); the "female inflorescence."

Hair point Awn, usually formed by an excurrent costa.

Haplolepidous Form of arthrodontous peristome; (**originally**) having main teeth formed from a single column of cells up dorsal face; (**strictly**) having only one circle of teeth derived from thickening of the contiguous walls of the primary and inner peristomial layers [homologous to the inner diplolepidous circle, endostome].

Heterophyllous Having leaves of different sizes or shapes.

Homomallous Pointing the same way.

Humus Decomposing organic material.

Hyaline Colorless or transparent.

Imbricate Closely appressed and overlapping; e.g., with the leaf margins overlapping like shingles on a roof.

Immersed Submerged or below the surface; referring to a capsule or perianth exceeded by the blades or awns of the perichaetial leaves.

Inclined Bent down; capsules that are between the erect and horizontal positions; i.e., drooping.

Incrassate With thickened cell walls.

Incurved Curved upward and inward, subjectively stronger than inflexed and weaker than involute; applied to leaf tips and margins.

Inoperculate Lacking an operculum.

Insertion The place or line of attachment of a structure; applied to leaves and branch on a stem, peristome, etc.

Isodiametric About as broad as long; applied to cells about the same diameter in all directions; including square, rounded or hexagonal.

Isophyllous Having similar stem and branch leaves.

Isosporous Having spores with the same distribution of size.

Julaceus Smoothly cylindrical, like a catkin, referring to stems or branches with strongly imbricate leaves.

Juxtacostal Region adjacent to costa.

Keeled Sharply folded along the middle, like the keel of a boat; V-shaped in cross-section.

Laciniate Fringed with lacinia; subjectively stronger than fimbriate, but not necessarily formed by cell erosion.

Lamina The flattened, generally unistratose, green part of the leaf blade excluding the costa and border.

Lanceolate Lance-shaped, in bryology narrow and tapered from near the base; narrowly ovate-acuminate.

Lateral At the side; e.g., lateral branches.

Lax Loose; referring to large thin-walled cells, as well as to nature and spacing of leaves on stem, or of stems in a tuft; laxly.

Leucocysts Large, empty, water-storage cells as in leaves of *Sphagnum* and *Leucobryum*.

Lignicolous Growing on logs.

Ligulate Strap-shaped; narrow, moderately long, sides parallel.

Limbate Bordered by distinct, elongate, incrassate marginal cells; e.g., *Fissidens*, *Syrrhopodon*.

Limbidium Border; differentiated leaf margin.

Linear Very narrow, elongate with nearly parallel sides; narrower than ligulate.

Lingulate Tongue-shaped; oblong with a slightly broadened apex.

Lumen The cell cavity.

Mammillate Convex to hemispherical with a blunt central projection.

Microphyllous Small-leaved; microphyllous branches produce smaller than normal leaves (cf. flagelliform).

Mitrate Conical and undivided or equally lobed at base, referring to a calyptra.

Mucronate Ending abruptly in a short point, usually caused by a shortly excurrent costa (**Apiculate** is somewhat longer; **Cuspidate** is even longer and stouter).

Naked Lacking covering structures or ornamentation; e.g., without hairs or papillae, referring to smooth, glabrous calyptra; without perichaetial leaves, referring to solitary archegonia borne in leaf axils.

Neck The sterile basal portion of a capsule, sometimes considerably differentiated; e.g., *Trematodon*; also upper narrow part of archegonium.

Oblanceolate Widest well above midlength; inversely lanceolate.

Oblate Wider than long.

Oblique Slanted; e.g., an oblique leaf insertion is one that is between transverse and longitudinal.

Oblong Rectangular with rounded corners or ends.

Obovate Egg-shaped with apex broader than base.

Obovoid Egg-shaped with the narrow end outward

Obtuse Broadly pointed; blunt or rounded when referring to apical region of a leaf.

Operculum The lid covering the mouth of most moss capsules; usually separated from the mouth by an annulus to open the capsule.

Ovate Outline of an egg with base broader than apex.

Ovoid Egg-shaped (in three dimensions).

Ovum Female gamete or egg, situated in the venter of the archegonium.

Papillose Bearing papillae; **unipapillose** - bearing one simple unbranched papilla on the cell surface;

multipapillose (**pluripapillose**) - bearing several papillae, or one compound or branched papilla on the cell surface. Loosely applied to any minutely rough surface that may be strictly mammillose.

Páramo A neotropical ecosystem located at high elevations between the upper forest line (about 3100 m altitude) and the permanent snow line (about 5000 m); dominated by herbs, grasses, and shurbs.

Paraphyllium Small green outgrowths of various shapes, i.e., filiform, lanceolate, scale- or leaf-like, or sometimes branched; produced randomly on the stems or branches of many pleurocarpous mosses; e.g., *Thuidium*.

Patent Leaves spreading from stem at an angle of 45° or more.

Pellucid Clear, translucent, or transparent.

Pendent Hanging downward (cf. inclined, pendulous).

Percurrent Extending to the apex.

Perforate Pierced through.

Perichaetial leaf Modified leaf or underleaf (bract; bracteole) associated with the gynoecium; collectively forming the perichaetium.

Perichaetium The gynoecium; strictly the ensheathing cluster of modified leaves or underleaves (bracts; bracteoles) and perianth, if present, enclosing the archegonia.

Perigonial leaf Modified leaf or underleaf (bract; bracteole) associated with the androecium; collectively forming the perigonium.

Perigonium The androecium; strictly the cluster of modified leaves or underleaves (bracts; bracteoles) enclosing the antheridia.

Peristome A circular structure, generally divided into 2n (i.e., 4, 8, l6, 32, or 64) teeth, arranged in a single or double (rarely multiple) row around the mouth of a capsule.

Piliferous With a hair point.

Pilose With long hairs; e.g., calyptra of some *Orthotrichum* species.

Pinnate With numerous spreading branches on opposite sides of the axis and thus resembling a feather.

Pitted Having small depressions or holes in the cell wall; sometimes called pores; in chlorocysts they are conspicuous as depressions in thickened walls between adjacent cells.

Plane Flat, not curved or wavy, referring to leaf margins or blade.

Pleurocarpous Producing sporophytes laterally from a perichaetial bud or a short specialized branch rather than at the stem tip (as in **Acrocarpous** mosses); with stems usually prostrate, creeping, freely branched mosses growing in mats rather than tufts.

Plicate With longitudinal furrows or pleats.

Porate With rounded apertures only.

Porose Having pores.

Proboscis Specialized leaf or leaf extension consisting mainly of costa, sometimes bearing gemmae; in Calymperaceae.

Propagulum Reduced bud, branch, or leaf serving in vegetative reproduction.

Prorulose With cell end(s) projecting slightly upward from lamina.

Prosenchymatous Made up of narrow elongate cells with tapered overlapping end walls.

Prostome A rudimentary structure outside, and usually adhering to, the main peristome teeth.

Prostrate Laying flat on the ground; creeping.

Pseudopodium An elongation of the gametophyte axis below the sporophyte in *Sphagnum* and *Andreaea*, serving the function of a seta; also applied to a similar extension of a stem tip bearing clusters of gemmae.

Puna An alpine biological community in the central portion of the Andes Mountains characterized by low growing, widely spaced plants.

Pyriform Pear-shaped; e.g., capsules of *Bryum*.

Radiculose Covered with rhizoids.

Ramentose Covered with flattened, multiseriate, scale-like structures in the perichaetium and on the calyptra, as in *Neckeropsis*.

Ramicolous Growing on bark.

Recurved Curved downward and inward; in leaves, referring to margins, apices, or marginal teeth; in peristome teeth curved outward and more or less downward.

Reflexed Bent down and inward generally referring to leaf margins or leaves on a stem.

Reniform Kidney-shaped.

Reticulate Netted, as the network pattern produced by cell wall thickenings.

Revoluble Rolling away, referring to an annulus that falls in a broken ring.

Revolute Rolled downward and backward, referring to a leaf margin.

Rheophilous Thriving in running water, dwelling in flowing creeks.

Rhizoids Hair-like structure that function in absorption and anchorage; in liverworts and hornworts one-celled and usually hyaline; in mosses usually brown to reddish, simple or branched, multicellular filaments, generally with oblique end-walls.

Rostrate Beaked, narrowed into a slender tip or point.

Rostrum Beak.

Rosulate Collected into a rosette; e.g., with the habit of small apparently acaulescent plants in which basal bracts grade continuously into about six, more or less uniform, broad spreading, vegetative leaves.

Rudimentary Incompletely developed, vestigial.

Rugose With transverse wrinkles or undulations; e.g., leaves of *Neckera*.

Saxicolous Growing on rocks.

Scabrous Rough to the touch, like sandpaper.

Secund Turned to one side; e.g., leaves on a stem.

Segments A piece cut off; a division of a plant organ such as one of the two parts of a bifid leaf; also applied to a cell derived from or cut directly from an apical cell; the main division of the diplolepidous endostome.

Septate Divided by cell walls; having partitions.

Seriate In rows; e.g., uni-, bi-, tri-, multiseriate; applied either to adjacent rows of leaf cells, or to ranks of leaves on a stem.

Serrate Saw-toothed; with marginal teeth pointing upward (towards apex).

Serrulate Minutely serrate.

Sessile Without a stalk or seta.

Seta Wlongated portion of the sporophyte between the capsule and foot (cf. pseudopodium).

Setaceous Bristle-like.

Shouldered Having an area of abrupt narrowing; e.g., the area on a leaf where the leaf base is abruptly narrowed to the upper lamina or blade or a similar constriction on an exostome tooth.

Spathulate Tapering proximally from a broad, rounded apex.

Spiculose Sharply, minutely toothed or papillose.

Spinose With sharp, pointed teeth; also very high, sharp leaf cell papillae or mammillae.

Spore A reproductive unit produced in the capsule as a result of meiosis; usually minute, mostly spherical and generally unicellular bodies that give rise on germination to a protonemata.

Sporophyte The spore-bearing generation; initiated by the fertilization of an egg; remaining attached to the gametophyte and partially dependent on it; typically consisting of foot, seta, and capsule.

Spreading Forming an angle of 45° or more.

Squarrose Spreading at right angles.

Stegocarpous Referring to capsules with a dehiscent operculum.

Stem The main gametophyte axis of mosses and leafy liverworts; grows by means of a single apical cell.

Stipe Leaves from the unbranched basal part of an erect stem in a dendroid or frondose moss (Pterobryaceae).

Stipitate Having a stipe or special stalk-like base.

Stoloniferous With slender, elongate branches possessing reduced or vestigial leaves throughout (cf. flagelliform).

Striate Marked with fine ridges or lines.
Striolate Finely ridged.
Strumose With a goiter-like swelling (or struma) at one side of the base, applied to some capsules; e.g., *Trematodon*.
Subula A long, slender point.
Subulate Slenderly long-acuminate.
Symmetrical Capable of division by a longitudinal plane into similar halves.
Teniola Intramarginal row (border) of differentiated cells in a leaf; generally more than one cell thick in the blade, always so in the base; e.g., *Calymperes*.
Terete Rounded in cross-section.
Terricolous Growing on soil (terrestrial).
Tomentose Woolly, densely radiculose; beset with tomentum.
Tomentum A felt-like covering of abundant rhizoids, on some stems or rarely leaves.
Trabecula Projecting cross-bars formed from horizontal walls on either face of arthrodontous peristome teeth; also strands of cells bridging air spaces within some capsules.
Trabeculate Cross-barred.
Transverse Across; perpendicular to the long axis of a plant or organ; e.g., transverse leaf insertion.
Truncate Abruptly cut off or squared, such as at the apex.
Tuberculate With peg-like projections.
Tufts Growth form with stems erect but radiating at the edges; small cushions; caespitose habit.
Undulate Wavy, such as the margin of a leaf.
Unipapillose With a single papilla per cell.
Unistratose One-layered; comprised of a single cell layer; e.g., most Bryophyte leaves.
Urn Spore-bearing portion of a capsule.
Urceolate Urn-shaped; applied to capsules constricted below a wide mouth and abruptly narrowed to the seta.
Vaginant lamina In *Fissidens*, one of the two clasping laminae below the apical lamina; one of the equitant laminae at base of the leaf that clasp the stem and enclose the base of the leaf above it.
Vaginula A ring or sheath enveloping the base of the seta, derived from the base of the archegonium and surrounding stem tissue and remaining after the separation of the calyptra.
Venter The swollen basal portion of an archegonium, containing the ovum.
Ventral (Of leaves) the top, or upper surface; (of peristome teeth) the inner face; (of stems or thalli) the lower surface, next to the substrate.
Vermicular Worm-shaped; long, narrow, somewhat wavy, commonly with rounded ends; usually applied to cells.
Vestigial Not fully developed.

REFERENCES

Allen, B.H. 1994. Moss flora of Central America. Part 1. Sphagnaceae-Calymperaceae. Monogr. Syst. Bot. Missouri Bot. Gard. 49:1–242.

Allen, B.H. 2002. Moss flora of Central America. Part 2. Encalyptaceae-Orthotrichaceae. Monogr. Syst. Bot. Missouri Bot. Gard. 90:1–699.

Brako, L. and J.L. Zarucchi. 1993. Catalogue of the flowering plants and gymnosperms of Peru. Missouri Botanical Garden, St. Louis, USA.

Buck, W.R. 1994. A new attempt at understanding the Meteoriaceae. J. Hattori Bot. Lab. 75:51–72.

Buck, W.R. 1998. Pleurocarpous mosses of the West Indies. Mem. New York Bot. Gard. 82:1–400.

Buck, W.R. 2002. Guide to the plants of Central French Guiana. Part 3. Mosses. Mem. New York Bot. Gard. 76:1–167.

Churchill, S.P. 1994. The mosses of Amazonian Ecuador. AAU Rep. 35:1–211.

Churchill, S.P. 1998. Catalog of Amazonian mosses. J. Hattori Bot. Lab. 85:191–238.

Churchill, S.P., I. Sastre-De Jesús, and H. Balslev. 1992. The mosses of Añangu, Napo Province, Ecuador. Lindbergia 17:50–54.

Churchill, S.P. and E. Linares C. 1995. Prodromus Bryologiae Novo-Granatensis Introduccion a la flora de musgos de Colombia.

Cornejo, F. and J.P. Janovec. 2010. A color guide to seeds of Amazonian plants. Princeton University Press, Princeton, NJ.

Croat, T.B. 1978. Flora of Barro Colorado Island. Stanford University Press, CA.

Crum, H. and L.E. Anderson. 1981. Mosses of Eastern North America. Columbia University Press, New York, NY.

Davis, S.D., V.H. Heywood, O. Herrera-MacBryde, J. Villa-Lobos, and A.C. Hamilton. 1997. The Americas. Volume 3. In: V.H. Heywood and S.D. Davis, eds. Centres of plant diversity, a guide and strategy for their conservation. World Wildlife Fund for Nature (WWF) and IUCN- The World Conservation Union.

Fleischer, M. 1861. Die Musci der Flora von Buitenzorg: zugleich Laubmoosflora von Java / bearb. von Max Fleischer. Leiden: E.J. Brill, 1900–1922.

Foster, R., T.A. Parker III, A. Gentry, L.H. Emmons, A. Chicchón, T. Schulenberg, L. Rodríguez, G. Lamas, H. Ortega, J. Icochea, W. Wust, M. Romo, J. Alban Castillo, O. Phillips, C. Reynel, A. Kratter, P.K. Donahue, and J.J. Barkeley. 1994. The Tambopata-Candamo Reserved Zone of southeastern Perú: a biological assessment. RAP Working Papers, Vol 6. Conservation International.

Foster, R.B. 1990. The floristic composition of the Rio Manu floodplain forest. In: A.H. Gentry, ed. Four neotropical rainforests. Yale University Press, New Haven, CT. Pp. 99–111.

Frodin, D.G. 2001. Guide to standard floras of the world, Edition 2. Cambridge University Press, Cambridge, England.

Gentry, A.H. 1982a. Neotropical floristic diversity: phytogeographical connections between Central and South America, Pleistocene climatic fluctuations, or an accident of the Andean Orogeny? Ann. Missouri Bot. Gard. 69:557–593.

Gentry, A.H. 1982b. Patterns of neotropical plant species diversity. Evol. Biol. 15:1–84.

Gentry, A.H. and B. León. 1997. Tambopata region, Peru. In: S.D. Davis, V.H. Heywood, O. Herrera-MacBryde, J. Villa-Lobos, and A.C. Hamilton, eds. Centres of plant diversity: a guide and strategy for their conservation, Volume 3—The Americas. WWF and IUCN, Oxford, England. Pp. 355–259.

Gentry, A. 1994. Importance of the Explorer's Inn Reserve permanent plots. In: R.B. Foster, J.L. Carr, and A.B. Forsyth, eds. The Tambopata-Candamo Reserved Zone of southeastern Peru: a biological assessment. Conservation International, Washington, DC. Pp. 55–57.

Ignatov, M.S. and S. Huttunen. 2002. Brachytheciaceae (Bryophyta)—a family of sibling genera. Arctoa 11:245–296.

Leigh, E.G. 1999. Tropical foresty ecology: a view from Barro Colorado Island. Oxford University Press, New York, NY.

Majestyk, P. and J. Janovec. 2004. Contributions to the moss flora of the Amazonian lowlands of Madre de Dios, Peru. Sida 21:1231–1238.

Menzel, M. 1992. Preliminary checklist of the mosses of Peru. J. Hattori Bot. Lab. 71:175–254.

Mori, S.A., G. Cremers, C. Gracie, J.-J. de Granville, M. Hoff, and J. D. Mitchell. 1997. Guide to the vascular plants of central French Guiana. Part 1. Pteridophytes, gymnosperms, and monocotyledons. Mem. New York Bot. Gard. 76:1–422.

Mori, S.A., G. Cremers, C. Gracie, J.-J. de Granville, S. V. Heald, M. Hoff, and J. D. Mitchell. 2002. Guide to the vascular plants of central French Guiana. Part 2. Dicotyledons. Mem. New York Bot. Gard. 76(2):1–776.

Phillips, O. and A.H. Gentry. 1993a. The useful plants of Tambopata, Peru: I. Statistical hypotheses tests with a new quantitative technique. Econ. Bot. 47:15–32.

Phillips, O. and A.H. Gentry. 1993b. The useful plants of Tambopata, Peru: II. Additional hypotheses testing in quantitative ethnobotany. Econ. Bot. 47:15–32.

Phillips, O.L., P. Núñez Vargas, A. Lorenzo Monteagudo, A. Peña Cruz, M.-E. Chuspe Zans, W. Galiano Sánchez, M. Yli-Halla, and S. Rose. 2003a. Habitat association among Amazonian tree species: a landscape-scale approach. J. Ecol. 91:757–775.

Phillips, O.L., R. Vásquez Martínez, P. Núñez Vargas, A. Lorenzo Monteagudo, M.-E. Chuspe Zans, W. Galiano Sánchez, A. Peña Cruz, M. Timaná, M. Yli-Halla, and S. Rose. 2003b. Efficient plot-based tropical forest floristic assessment. J. Trop. Ecol. 19:629–645.

Pitman, N.C.A., et al. 1999. Tree species distributions in an upper Amazonian forest. Ecology 80:2651–2661.

Pitman, N.C.A., et al. 2001. Dominance and distribution of tree species in upper Amazonian terra firme forests. Ecology 82:2101–2117.

Ribeiro, J.E.L.S., M.J.G. Hopkins, A. Vicentini, C.A. Sothers, M.A.S. Costa, J.M. Brito, M.A.D. da Souza, L.H.P. Martins, L.G. Lohmann, P.A.C.L. Assuncao, E.C. Pereira, C.F. Silva, M.R. Mesquita, and L.C. Procopio. 1999. Flora da Reserva Ducke: guia de identificacao de uma floresta de terra-firme na Amazonia Central. INPA/DIFD, Manaus, Brazil.

Sharp, A.J., H. Crum, and P. Eckel (eds.) 1994. The moss flora of Mexico. Mem. New York Bot. Gard. 69:1–1113.

Smith, A., B. Leon, H. Tuomisto, H. Werff, R. Moran, M. Lehnert, and M. Kessler. 2005. New records of pteridophytes for the flora of Peru. Sida 21:2321–2342.

Stewart, P.D. 1988. Tambopata Reserved Zone, southeast Peru. Oryx 22:95–99.

Terborgh, J. 1994. An overview of research at the Cocha Cashu Biological Station. In: A.H. Gentry, ed. Four neotropical rainforests. Yale University Press, New Haven and London. Pp. 48–59.

Terborgh, J. and E. Anderson. 1998. The composition of Amazonian forests: patterns of local and regional scales. J. Trop. Ecol. 14:645–664.

van der Werff, H. and T. Consiglio. 2004. Distribution and conservation significance of endemic species of flowering plants in Peru. Biodivers. Conservation 13:1699–1713.

Vargas, C. 1974. La flora del departamento de Madre de Dios (Perú). Lima, Perú.

Vasquez, R. A.R. Lleras, and C.M. Taylor. 2001. Florula de las Reservas Biologicas de Iquitos, Peru. Missouri Botanical Garden Press, St. Louis, MO.

INDEX

Accepted generic names are listed in **bold**, synonyms or occasionally used names in *italics*, and families in ALL CAPS.

Alphabetical List of Families with Genera